THE
EINSTEIN
Scrapbook

THE EINSTEIN *Scrapbook*

ZE'EV ROSENKRANZ

THE ALBERT EINSTEIN ARCHIVES
THE JEWISH NATIONAL & UNIVERSITY LIBRARY
THE HEBREW UNIVERSITY OF JERUSALEM

THE JOHNS HOPKINS UNIVERSITY PRESS
BALTIMORE AND LONDON

1928

© 1998, 2002 The Jewish National & University
Library, Jerusalem

All rights reserved. Published 2002

Printed in the United States of America on acid-free paper
9 8 7 6 5 4 3 2

The Johns Hopkins University Press
2715 North Charles Street
Baltimore, Maryland 21218-4363
www.press.jhu.edu

Designed by Jody Billert, Design Literate, Inc.

An earlier version appeared in a limited edition by
the Jewish National & University Library, Jerusalem,
in 1998 under the title *Albert through the Looking Glass.*
www.albert-einstein.org

Library of Congress Cataloging-in-Publication Data

Rosenkranz, Ze'ev
The Einstein scrapbook / Ze'ev Rosenkranz.
     p.   cm
Includes bibliographical references and index.
ISBN 0-8018-7203-0 (hardcover : alk. paper)
1. Einstein, Albert, 1879–1955. 2. Einstein, Albert, 1879–1955—
Archives. I. Bet ha-sefarim ha-le'umi veha-universita'i
bi Yerushalayim. Albert Einstein Archives. II. Title.
QC 16.E5 R674 2002
530'.092—dc21   2002005379

A catalog record for this book is available from the British Library.

IN LOVING MEMORY

OF MY FATHER,

ARNOLD ROSENKRANZ

(1923–1999)

Contents

Foreword

In these pages, the reader is given a view, admittedly limited, of one of the most universal symbols of the twentieth century, one of the greatest intellects of all ages. Albert Einstein has become a myth, a symbol, a paradigm of scientific revolution. But the personal papers of Einstein which are stored at The Hebrew University's Jewish National & University Library reveal that there is much that is hidden behind the myth.

Paradoxically, Einstein's image as reflected by this book is far from two-dimensional. Einstein's theory of relativity constitutes one of the most profound revolutions in the human perception of the world in which we live, and represents one of the most significant contributions to human knowledge. Another revolution in scientific thought, arguably even more profound, is that of quantum mechanics. Einstein's ongoing struggle with the theory of quantum mechanics increasingly isolated him from the mainstream of scientific thought and could have turned him into a tragic figure. But despite this, none of the creators of quantum mechanics has been as mythified in the public eye as Einstein.

What was so special about the figure of Einstein that it captured the collective imagination of the twentieth century? Our Einstein Archives provides some of the answers. It reveals a complex, multifaceted person, a man of contradictions and paradoxes. This brilliant intellect, unfettered by the chains of existing paradigms, created through his special and general theories of relativity one of the most far-reaching breakthroughs in the history of human thought. At the same time he was so adamant in his convictions that he was unwilling to accept the other revolution in twentieth-century science, the theory of quantum mechanics.

Although a dedicated pacifist, Einstein felt compelled to advocate the development of nuclear weapons in light of the threat of a Nazi bomb emerging as part of the horrible destructive capacity of the German war machine.

Although he did not believe in the concept of a personal God, he was a life-long atheist deeply influenced by his Jewish identity, by a concept of peoplehood and the unique role of the Jewish people in the unfolding saga of human civilization.

Even in his relations with The Hebrew University one finds the same contradictions. One of its founding fathers, Einstein was at the same time one of its most severe critics joining with Chaim Weizmann in challenging the lines of development and leadership expressed by Judah L. Magnes, the then head of the institution. Despite this, he chose The Hebrew University as the permanent repository for his personal and scientific manuscripts, papers, and correspondence.

I believe that these tensions reveal the complexity of the real Albert Einstein and explain perhaps, at least in part, why he and no other person became such a symbol of the twentieth century.

PROF. MENACHEM MAGIDOR
PRESIDENT
THE HEBREW UNIVERSITY OF JERUSALEM

Preface

This book offers a unique glimpse into the life and work of Albert Einstein, one of the most prominent and influential figures of the modern era. Based on a selection of fascinating documents and photographs from his personal papers, it provides us with a kaleidoscope—a looking glass—through which to perceive this great scientist, humanist, and Jew. Some items are published here for the first time.

Einstein's papers reflect the complexity of his character and the exceptional quality of his genius. The repository of these papers is the Albert Einstein Archives, housed in the Department of Manuscripts & Archives of The Jewish National & University Library (JNUL) at The Hebrew University of Jerusalem. Einstein was one of the founders of The Hebrew University, and his special relationship to this institution found a lasting expression in the bequest of his literary estate and personal papers to the university in his Last Will and Testament.

As a preeminent physicist, Einstein radically transformed our understanding of the universe. As an ardent humanist, he took an active and outspoken stand on the significant political and social issues of his time. And, as a committed Jew, he advocated a distinctive moral role for the Jewish people.

This volume displays the breadth and diversity of Einstein's personality, his scientific and other intellectual pursuits, as well as his political and social endeavors. Each chapter examines a different dimension of Einstein's life and work, and begins with a short introductory passage that provides a general thematic overview. Reproductions of relevant documents and photographs illustrate particular aspects of each topic, and are accompanied by short descriptions. English translations of manuscripts and letters are provided where necessary.

In selecting the archival material reproduced here, I have been guided by my familiarity with and knowledge of the holdings of the Albert Einstein Archives. In addition, I drew upon lists of select items compiled by Einstein's personal secretary, the late Helen Dukas, in the 1960s. Of course, I was able to include in this limited space only a tiny portion of the Archives' holdings. Ultimately, I sought a wide variety of materials in order to represent the scope of the private and public selves of Albert Einstein.

The book opens with a timeline of the major events in Einstein's life. An engaging look at Einstein's biography follows, beginning with an exploration of his family background and education and then offering fascinating insights into his intriguing personality, his complex marital relations, and his role as a father. The book proceeds with a detailed presentation of Einstein's scientific theories, which changed the face of modern physics and earned him both world renown and an enduring place in history. This section traces the trajectory of Einstein's scientific career through its various stages: from his breakthrough articles on the special theory of relativity, quantum theory, and Brownian motion published in 1905—his *annus mirabilis*—through his revolutionary general theory of relativity in 1916, his being awarded the Nobel Prize for Physics for the year 1921, and his opposition to quantum mechanics from the mid-1920s. It concludes with his final quest for a unified field theory. Turning then to his active commitments outside the scientific sphere, Einstein's role in the international peace movement and his views on social and economic affairs are discussed. In the next chapter, Einstein's unique relationship to the Jewish people is considered in a section devoted to his Jewish identity. A discussion of the years he spent in the United States follows. Then, on the lighter side,

Einstein's passion for music and sailing, his charming correspondence with children from around the world, and some of the quirky letters he received are covered in the next chapters. The work concludes with an analysis of Einstein's mythic status—the process by which he became a universal cultural icon. Last, positioned at the end of the volume, is a profile of the Albert Einstein Archives itself, and also a detailed bibliography and index.

It gives me great pleasure to express my gratitude to those who have made this book possible.

First, I would like to thank the numerous persons and institutions that have kindly granted us permission to include here the documents, photographs, and cartoons to which they hold the copyright. They are credited in the volume alongside the items which they have let us reproduce. Particular thanks are due to Princeton University Press for their kind cooperation. In preparing this publication, I have consulted a considerable number of books and articles by and on Albert Einstein. Details of these can be found in the bibliography.

Next, I would like to single out a number of individuals for their assistance in the preparation of this book for publication. I wish to thank Astrid Gottwald, Susan Worthington, and Judith Levy for their archival research; Barbara Wolff for her photo research; and Jeffrey Mandl for his substantial help in editing the manuscript of this volume. Gratitude is also due to Esther Guggenheim for her work on the material pertaining to Einstein and the United States, to Aurelie Segal for supervising the scanning of the documents, to Hannah Katzenstein for proofing the manuscript, and to the late

Dina Carter for assisting with the publication. My thanks to Racheline G. Habousha, Reuven Koffler, and Prof. Ya'akov Wahrman, who kindly provided background information on several of the photographs included here. Further, Dr. Alfredo T. Tolmasquim, Dr. Anthony Travis, and Prof. Issachar Unna are to be particularly thanked for sharing their scholarly knowledge with me as work proceeded on the project.

I am especially indebted to Eliyahu Honig, associate vice-president of The Hebrew University, and The Victorian Friends of The Hebrew University in Australia, who were instrumental in raising the necessary funds for the 1998 version of this publication. Furthermore, I am grateful to the Trustees of the Estate of Miriam and Israel Blankfield and other donors for their generous financial contributions to this project. Special thanks are due to Prof. Israel Shatzman and to Prof. Sara Japhet, former directors of the JNUL, for their whole-hearted support of this project. I am also grateful to Rafael Weiser, director of the Department of Manuscripts and Archives at the Library, and to his staff for their helpful advice and assistance.

Finally, I would like to express my personal gratitude to my wife, Brenda Shorkend, who not only helped edit the text but also provided me with great encouragement and moral support.

ZE'EV ROSENKRANZ
BERN DIBNER CURATOR
ALBERT EINSTEIN ARCHIVES

THE
EINSTEIN
Scrapbook

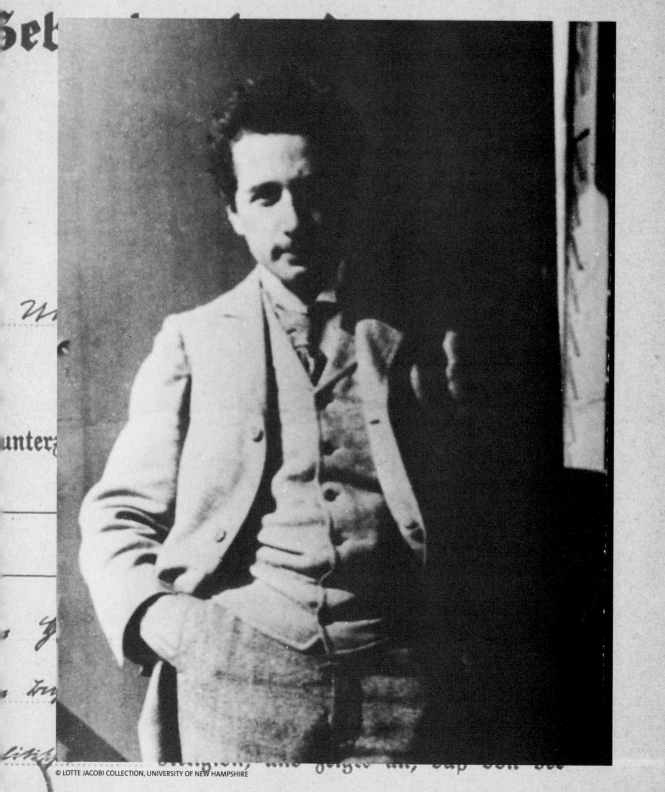

1 | The Einstein Timeline

Earliest known photo of Albert Einstein

Studio portrait at age fourteen

1898

The Early Years

1879 Born March 14 at 11:30 A.M. in Ulm, Germany

1880 Einstein family moves to Munich

1885–88 Pupil at Catholic elementary school in Munich • Private lessons in Judaism at home

1888–94 Pupil at Luitpold-Gymnasium, Munich • Religious instruction at school (until 1892)

1894 Parents move to Milan • Six months later, Einstein leaves Gymnasium without completing his schooling and joins his family in Pavia, Italy

The Swiss Years

1895–96 Pupil at cantonal school in Aarau, Switzerland

1896 Renounces his German citizenship

1896– Student at the Polytechnic (later the Federal
1900 Institute of Technology), Zurich

1901 Acquires Swiss citizenship • Completes his first scientific paper

1901–2 Temporary teaching position at school in Schaffhausen, Switzerland

1902 Daughter Lieserl born to Mileva Marić • Appointed as technical expert third class at the Swiss Patent Office in Bern

1903 Marriage to Mileva Marić in Bern • Founds "Akademie Olympia" with Conrad Habicht and Maurice Solovine • Daughter Lieserl probably put up for adoption

1904 Son Hans Albert born in Bern

1905 The *annus mirabilis:* completes papers on light quanta, Brownian motion, and special theory of relativity • Receives Ph.D. from Zurich University

1906 Promoted to technical expert second class at the Swiss Patent Office

1907 Discovers the principle of equivalence

1908 Appointed lecturer at Bern University

1909 Resigns from Patent Office • Appointed associate professor of theoretical physics at Zurich University

1910 Second son Eduard born in Zurich

1911 Predicts bending of light

1911–12 Professor of theoretical physics at German University of Prague

1912–14 Professor of theoretical physics at the Federal Institute of Technology, Zurich

The Berlin Years

1914 Appointed professor at University of Berlin (without teaching obligations) and member of Prussian Academy of Sciences • Separates from his wife, Mileva Marić—she returns to Zurich with his two sons • Signs anti-war "Manifesto to Europeans"

1915 Joins pacifist "New Fatherland League" • Completes logical structure of the general theory of relativity

1916 Publication of the general theory of relativity

1917 Writes first paper on cosmology • Appointed director of Kaiser Wilhelm Institute for Physics in Berlin

1917–20 Suffers from a liver ailment, a stomach ulcer, jaundice, and general weakness—his cousin Elsa Loewenthal Einstein takes care of him

1918 Supports the new Weimar Republic in Germany

1919 Divorces Mileva Marić • Bending of light observed during solar eclipse in West Africa and Brazil • First discussions on Zionism with Kurt Blumenfeld • Marries his cousin Elsa • Announcement at joint meeting of Royal Society and Royal Astronomical Society that Einstein's theories have been confirmed by eclipse observations • Sensational headlines in the *Times* and the *New York Times*: Einstein becomes a world figure

1920 Mass meeting against the general theory of relativity in Berlin • Appointed special visiting professor at Leiden University

1921 First visit to the United States with Chaim Weizmann: fund-raising tour for The Hebrew University • Lectures at Princeton University on theory of relativity

c. 1905

1916

1928

1931

1934

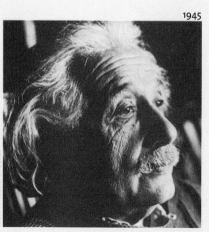

1945

1922 Completes first paper on unified field theory •
Visit to Paris contributes to normalization of
French-German relations • Joins Committee
on Intellectual Cooperation of the League of
Nations • Lecture tours in Japan and China •
Awarded Nobel Prize for Physics for 1921

1923 Visit to Palestine: holds inaugural scientific
lecture at future site of The Hebrew University
in Jerusalem; named first honorary citizen of
Tel Aviv • Visit to Spain • Lecture in
acknowledgment of Nobel Prize in Göteborg,
Sweden • Edits first collection of scientific
papers of The Hebrew University

1924 The "Einstein-Institute" in Potsdam, Germany,
housed in the "Einstein-Tower," starts its
activities

1925 Trip to South America: Argentina, Brazil,
and Uruguay • Signs manifesto against
obligatory military service • Joins Board
of Governors and Academic Council of The
Hebrew University

1927 Begins intense debate with Niels Bohr on the
foundations of quantum mechanics

1928 Suffers temporary physical collapse—enlarge-
ment of the heart is diagnosed

1930 Intensive activity on behalf of pacifism

1930–32 Three trips to United States: stays mainly at
the California Institute of Technology,
Pasadena, during winter semesters

1932 Supports conservation of the Weimar Republic
• Public correspondence with Sigmund Freud
on the nature of war • Appointed professor at
the Institute for Advanced Study, Princeton •
Plans to divide his time between Berlin and
Princeton • Leaves Germany for the last time

The Princeton Years

1933 Declares that he will not return to Germany •
Resigns from Prussian Academy of Sciences •
Spends spring and summer in Belgium and
Oxford • Emigrates to United States in
September • *Why War?* published

1934 Collection of essays *The World As I See It*
published

1935 The Einstein-Podolsky-Rosen paradox is
published

1936 Elsa Einstein dies

1938 Publication of *The Evolution of Physics*

1939 Signs famous letter to President Franklin D. Roosevelt recommending U.S. research on nuclear weapons

1940 Acquires U.S. citizenship

1943 Works as consultant with the Research and Development Division of the U.S. Navy Bureau of Ordnance, section Ammunition and Explosives

1944 Handwritten copy of his 1905 paper on special relativity auctioned for six million dollars in Kansas City, as a contribution to the American war effort

1945 Shattered by the extent of the Holocaust of European Jewry • Shocked by the nuclear bombing of Hiroshima and Nagasaki

1946 Becomes chairman of the Emergency Committee for Atomic Scientists • Expresses public support for the formation of a world government

1947 Intense activity on behalf of disarmament and world government

1948 Supports creation of the State of Israel • First wife, Mileva Marić, dies in Zurich • Intact aneurysm of the abdominal aorta disclosed

1949 Publication of "Autobiographical Notes"

1950 Signs Last Will and Testament: Otto Nathan and Helen Dukas named co-trustees • The Hebrew University named as the ultimate repository of his personal papers • Collection of essays, *Out of My Later Years,* published

1952 Offered presidency of the State of Israel

1953 Public support for individuals under investigation by the House Un-American Activities Committee

1955 Co-signs the Russell-Einstein Manifesto warning of the nuclear threat • Rupture of the aortic aneurysm • Dies April 18 at 1:15 A.M. in Princeton Hospital at the age of seventy-six • Body cremated and ashes scattered at an undisclosed place

1953

1954 (75th birthday)

1950s

Albert Einstein (five years old) with his sister Maja (three years old), 1884

2 | Einstein's Personal Life

"*I am truly a 'lone traveler'*
and have never belonged to
my country, my home,
my friends, or even my
immediate family, with my
whole heart; in the face
of all these ties, I have never
lost a sense of distance
and a need for solitude—
feelings which increase with
the years."

Geburtsurkunde.

Nr. 224

Ulm am 15. März 1879.

Vor dem unterzeichneten Standesbeamten erschien heute, der Persönlichkeit nach _____

_____ bekannt,

der *Kaufmann Hermann Einstein* _____

wohnhaft zu *Ulm Bahnhofstraße B. No 135* _____

_____ *israelitischer* Religion, und zeigte an, daß von der

Pauline Einstein geborenen *Koch, seine Ehefrau,* _____

_____ *israelitischer* Religion,

wohnhaft *bei ihm* _____

zu *Ulm in seiner Wohnung* _____

am _____ *vierzehn* ten *März* des Jahres

tausend acht hundert *sieben* zig und *neun vormittags*

um _____ *elf ein halb* _____ Uhr ein Kind *männ* lichen

Geschlechts geboren worden sei, welches _____ den _____ Vornamen

_____ *Albert* _____ erhalten habe.

_____

Earliest known photograph of
Albert Einstein

RIGHT Einstein's birth certificate, Ulm,
Germany, dated March 15, 1879

Family Background

Einstein's ancestors had settled in rural Swabia in southern Germany before 1750. His parents belonged to the assimilated, Jewish middle class. Einstein's father, Hermann Einstein, an unsuccessful merchant, was a kind-hearted, rather passive man who was fond of German literature. His mother, Pauline Einstein, née Koch, had a strong personality and was a talented pianist. Albert, their first child, was born in Ulm in 1879. Maja, his sister, was born two years later.

Studio portrait of Einstein age fourteen

Education

Contrary to popular belief, Einstein was a very good pupil. When he was a small child, Einstein's parents were concerned about his slow development. Yet, by the time he reached primary school, he had overcome his early difficulties.

Einstein attended a German high school but left at the age of sixteen, as he disliked the authoritarian, militaristic atmosphere he encountered there. He completed his school education in Aarau, Aargau Canton, Switzerland. His school-leaving certificate, dated October 3, 1896, includes the highest possible grade, which was six.

Der Erziehungsrat
des
Kantons Aargau

urkundet hiemit:

Herr *Albert Einstein* von *Ulm*,
geboren den 14. März 1879,

besuchte die aargauische Kantonsschule & zwar die III. & IV. Klasse der Gewerbeschule.

Nach abgelegter schriftl. & mündl. Maturitätsprüfung am 18., 19. & 21. September, sowie am 30. September 1896, erhielt derselbe folgende Noten:

1. Deutsche Sprache und Litteratur		5
2. Französische „ „ „		3
3. Englische „ „ „		—
4. Italienische „ „ „		5
5. Geschichte		6
6. Geographie		4
7. Algebra		6
8. Geometrie [Planimetrie, Trigonometrie, Stereometrie & analytische Geometrie]		6
9. Darstellende Geometrie		6
10. Physik		6
11. Chemie		5
12. Naturgeschichte		5
* 13. Im Kunstzeichnen		4
* 14. Im technischen Zeichnen		4

* Hier gelten die Jahresleistungen.

Gestützt hierauf wird demselben das Zeugnis der Reife erteilt.

Aarau, den 3ten Oktober 1896.

Im Namen des Erziehungsrates,
Der Präsident:

Der Sekretär:

The Board of Education of the Canton Aargau hereby certifies that: Mr. Albert Einstein of Ulm, born on March 14, 1879, attended the Aargau Kantons schule, namely the third and fourth classes of the technical school. After taking the written and the oral school-leaving examination held on September 18, 19, 21, and 30, 1896, he received the following marks:

German	5
French	3
English	—
Italian	5
History	6
Geography	4
Algebra	6
Geometry	6
Descriptive Geometry	6
Physics	6
Chemistry	5
Natural History	5
Artistic Drawing	4
Technical Drawing	4

Based on these marks, the above is granted the school-leaving certificate. Aarau, October 3, 1896

School-leaving certificate, Aarau, Aargau Canton, Switzerland, October 3, 1896

Mileva Marić and her two sons, Eduard and
Hans Albert, 1914

Albert Einstein with Mileva and Hans Albert,
ca. 1905–1906
PHOTO DONATED BY ELIZABETH EINSTEIN

Family Life

Albert Einstein had a turbulent private life. While
studying at the Zurich Polytechnic, he fell in love
with Mileva Marić, a fellow student who came from
Serbia. She was one of the first women to study
physics there. These first years of their relationship
were warm and passionate. Mileva gave birth to
their daughter, Lieserl, in 1902, yet because neither
of them had a secure income at the time, they did
not marry until a year later. All traces of Lieserl
after the age of two seem to have been lost. There
has been some speculation as to her fate—she may
have been put up for adoption or she may have died
at a young age. Albert and Mileva went on to have

Letter to his first wife, Mileva Marić, February 4, 1902

My dearest sweetheart,

Poor, dear sweetheart; what you've had to suffer if you can't even write to me yourself anymore! It's such a shame that our dear Lieserl must be introduced to the world this way! I hope you are bright and cheerful by the time my letter arrives. I was frightened out of my wits when your father's letter came because I had already sensed something wrong. All other fates are nothing compared to this. My first reaction was to remain a teacher at old N[üesch]'s for two more years if that could bring you health and happiness; but now you see that it really is a Lieserl, just as you'd wished. Is she healthy, and does she cry properly? What are her eyes like? Which one of us does she more resemble? Who is giving her milk? Is she hungry? She must be completely bald. I love her so much and don't even know her yet! Couldn't you have a photograph made of her in the meantime until you've regained your health? Is she looking at things yet? Now you can make observations. I'd like to make a Lieserl myself sometime—it must be fascinating! She's certainly able to cry already, but won't know how to laugh until much later. Therein lies a profound truth. When you feel a little better you'll have to draw a picture of her!

Dear Tete, *
When I read your letters I am very much reminded of my youth. In one's thoughts, one tends to set oneself against the world. One compares one's own strengths with everything else, one alternates between despondency and self-assurance. One has the feeling that life is eternal and that everything that one does and thinks is so important. Yes, one feels as if one were the first and only fellow who has gone through all this. Yet this heroism is rather embarrassing and can only be corrected by humor and by one's somehow turning with the social machine. All my life I have troubled myself with problems and am always—as on the first day—inspired by the fact that cognition in the scientific and artistic sense is the best thing we possess. My love of these things has never diminished and will stay with me till I breathe my last. You were also born for this and your words to the contrary only derive from the fear of not being able to achieve anything worthwhile. Dear Tetel,* therefore I somehow take pity on you. But there is an easy solution. One becomes a cog in the large machinery so that no one can demand anything else from one. One is a thinking and feeling creature privately and for one's own pleasure. If one hears the angels singing a couple of times during one's life, one can give the world something and one is a particularly fortunate and blessed individual. Yet if this is not the case, one is nevertheless a small particle of the soul of one's generation and that is also beautiful. Think about this carefully, so that you don't fall victim to the devil of ambition and vanity. And keep in mind: not the desire for the achievement but love of the things themselves can lead to something worthwhile. Be that as it may, you bring me great joy because you're not going through life mindlessly but rather seeing and thinking. I would like to be with you again soon. Couldn't you come here during your Easter holidays? (I don't dare to ask you to come during Christmas so that Mama will not be left all on her own.)

> —Letter to his younger son, Eduard Einstein,
> December 27, 1932.
> [*Nickname for Eduard]

Albert Einstein with his elder son, Hans Albert, and his grandson, Bernhard, 1932

Albert and Elsa Einstein on board the *Kitano Maru* during their voyage to the Far East, 1922

two sons: Hans Albert, born in 1904, and Eduard born in 1910.

During the years in which Einstein developed his early revolutionary theories, Mileva functioned as a sounding board for his ideas. Yet, there is no evidence that she actively participated in his scientific work. Apparently, Mileva never got over the loss of her daughter and increasingly suffered from depression. By around 1909 their relationship began to deteriorate. They became increasingly

Dear Elsa,

When we meet each other often in Berlin you will see that we will be good friends for life, who will be able to light up each other's existence. The most beautiful thing will be our walks in the Grunewald forest and when the weather is bad, our meetings in your small room. So you want to learn a big lecture by heart? Isn't that self-cruelty to animals? People like us are happy when we don't have to appear in public—and you do it without necessity. But your courage impresses me nevertheless. Yet, I wouldn't want to be there under any circumstances. Because I love the words that come from your mouth only as a private gift to me and a product of the moment, the hastier the better. If you were to recite the most beautiful poem ever so divinely, the joy I would derive from it would not come close to the joy I experienced when I received the mushrooms and goose cracklings you cooked. I know how the psychologist would interpret this but I wouldn't be ashamed of it and you would certainly not despise the primitive side of my nature which is thereby expressed, but you may smile at it. The extreme spiritualization of the mind has something solemnly oppressive about it and banishes joyful laughter.

Letter to his cousin and future wife, Elsa Loewenthal, November 7, 1913

EINSTEIN'S PERSONAL LIFE

estranged and in 1912, Albert became involved with his cousin, Elsa Loewenthal.

Upon Einstein's move to Berlin in 1914, he and Mileva separated. They were divorced in 1919, and soon afterwards he married Elsa, who had two daughters, Margot and Ilse, from a previous marriage. Albert's sons remained with Mileva in Zurich, yet he continued to be in contact with them. Einstein's relationship with Hans Albert was strained by the divorce and, in later years, he strongly disapproved of Hans Albert's marriage to an older woman. Einstein had a special affection for his younger son, Eduard, a highly sensitive and talented young man who eventually developed

Albert and Elsa Einstein at the Grand Canyon, February 28, 1931

PHOTO BY FRED HARVEY; COURTESY OF THE FRED HARVEY COMPANY, GRAND CANYON NATIONAL PARK

THE UNITED STATES OF AMERICA

CERTIFICATE OF NATURALIZATION

No. 5013865

Petition No. 4009

Personal description of holder as of date of naturalization: Age 61 years; sex Male; color White; complexion Medium; color of eyes Brown; color of hair Grey; height 5 feet 7 inches; weight 175 pounds; visible distinctive marks None; Marital status Widower; former nationality German.

I certify that the description above given is true, and that the photograph affixed hereto is a likeness of me.

Albert Einstein
(Complete and true signature of holder)

United States of America } ss:
District of New Jersey }

Be it known that ALBERT EINSTEIN then residing at 112 Mercer St., Princeton, New Jersey having petitioned to be admitted a citizen of the United States of America, and at a term of the DISTRICT Court of THE UNITED STATES held pursuant to law at Trenton, New Jersey on October 1st 19 40 the court having found that the petitioner intends to reside permanently in the United States, had in all respects complied with the Naturalization Laws of the United States in such case applicable, and was entitled to be so admitted, the court thereupon ordered that the petitioner be admitted as a citizen of the United States of America.

In testimony whereof the seal of the court is hereunto affixed this 1st day of October in the year of our Lord nineteen hundred and forty and of our Independence the one hundred and sixty-fifth.

Albert Einstein
seal

"This is a personal document and it is a breach of the U.S. Code (and punishable as such) to copy, print, photograph or otherwise illegally use it." See other side

Benjamin F. Havens
Clerk of the U.S. District Court.
By Hazel K. Fox Deputy Clerk.

DEPARTMENT OF LABOR

U.S. Certificate of Naturalization, Trenton, New Jersey, October 1, 1940

OPPOSITE Albert Einstein and his step-daughter, Margot, at the ceremony during which they received their U.S. citizenship, Trenton, New Jersey, October 1, 1940

PHOTO BY MARTIN D'ARCY, *EVENING TIMES* PHOTOGRAPHER

schizophrenia. Elsa was much less Einstein's intellectual match than Mileva, yet she enjoyed his fame and protected him from intrusions. Though he was not always the most faithful of husbands, he cared for her deeply and enjoyed the cozy atmosphere she created in their home. They moved to Princeton in 1933 with Einstein's secretary, Helen Dukas. Later they were joined by Einstein's sister, Maja, and Elsa's daughter, Margot. After Elsa's death in 1936, Einstein lived a quiet life, working at the Institute

EINSTEIN'S PERSONAL LIFE

STATE OF NEW JERSEY
OFFICE OF REGISTRAR OF VITAL STATISTICS

NO. 227

of .... **Princeton Borough, Mercer County**
City, Borough or Township and County

This is to Certify that the following is correctly copied from a record of Death in my office.

NAME OF DECEASED	PLACE OF DEATH	DATE OF DEATH		
Albert Einstein	Princeton Hospital	April 18, 1955		

SOCIAL SECURITY NUMBER	SEX	COLOR	MARITAL CONDITION	DATE OF BIRTH	AGE		
					YRS.	MOS.	DAYS
	Male	White	Widower	March 14, 1879	76	1	4

PLACE OF BIRTH	CAUSE OF DEATH
Ulm, Germany	Rupture of Arteriosclerotic Aneurysm of Abdominal Aorta.

SUPPLEMENTAL INFORMATION IF DEATH WAS DUE TO EXTERNAL CAUSES

ACCIDENT, SUICIDE OR HOMICIDE	DATE OF OCCURRENCE
SPECIFY	

WHERE DID INJURY OCCUR?

CITY OR TOWN	COUNTY	STATE

DID INJURY OCCUR IN OR ABOUT HOME, ON FARM, IN INDUSTRIAL PLACE, IN PUBLIC PLACE?

	SPECIFY TYPE OF PLACE
WHILE AT WORK?	MEANS OF INJURY

NAME OF PERSON WHO CERTIFIED CAUSE OF DEATH	ADDRESS
Guy K. Dean, Jr., M. D.	Plainsboro, N. J.

David T. Blake
Registrar of Vital Statistics

.... Borough Hall, Princeton, N. J. ....
Address

.... April 26, ... 19 55
Date of Issue

B H 501NA

USHER PUBLISHING CO INC TRENTON N J

for Advanced Study and taking long summer vacations with family and friends. Einstein died of a rupture of an aortic aneurysm in 1955 at the age of seventy-six. He was cremated and his ashes scattered at an undisclosed place.

Death certificate, Princeton, New Jersey, April 26, 1955

OPPOSITE Albert Einstein with his secretary, Helen Dukas, holding leash for Chico, the dog, on a Princeton street, early 1950s

Einstein sticking out his tongue on his seventy-second birthday, March 14, 1951, with Mr. and Mrs. Frank Aydelotte. He had been asked for a "birthday pose."

Personality

Einstein was a solitary, dreamy child. By adolescence, he had evolved into a self-confident, even brash young man. He established a few close friendships with fellow students and clearly exuded a great deal of attraction for the opposite sex. During their courtship, he and his future wife Mileva rejected middle-class values and wanted to create a bohemian existence for themselves. Einstein was obviously capable of a deep relationship with Mileva, yet as a consequence of their failed marriage, he seems to have lowered his expectations for marital bliss. He never felt as deeply for his second wife, Elsa, but he enjoyed the

Albert Einstein with Helen Dukas and Chico, the dog, in their Princeton garden

reversion to a secure bourgeois existence, which his relationship with her provided. During their marriage, he sought intimacy in a number of extramarital affairs.

Throughout his life, Einstein was a radical nonconformist who rejected societal norms. He reserved a special sarcasm for every form of pomp and circumstance. In addition, he had a strongly developed sense of humor and especially liked Jewish jokes.

In his later years, Einstein distanced himself from his feelings and belittled the importance of emotional issues. He claimed that this detachment had been a necessary prerequisite for his total dedication to his first love—science. He stated that his preferred profession was to be a lighthouse keeper.

I regret that I cannot accede to your request, because I should like very much to remain in the darkness of not having been analyzed.
—Albert Einstein's response to the suggestion that he undergo Adlerian psychotherapy, January 1927. The German psychotherapist and left-wing politician, H. Freund, planned to carry out a study of politicians based on psychotherapy and asked Einstein to participate.

3 | Einstein's Scientific Achievements

"When the blind beetle crawls

over the surface of a globe,

he doesn't realize that the track

he has covered is curved. I was

lucky enough to have spotted it."

Einstein's Significance

Albert Einstein's contribution to modern physics is simply unique. His scientific career was a constant quest for the universal and immutable laws which govern the physical world. His theories spanned the fundamental questions of nature, from the very large to the very small, from the cosmos to sub-atomic particles. He overturned the established concepts of time and space, energy and matter. Einstein played a crucial role in establishing the two pillars of twentieth-century physics: he was the father of the theory of relativity and a major contributor to quantum theory. Einstein was a theoretical physicist—his only concrete tools being pencil and paper. It has been said that his true tools were a penetrating and intuitive grasp of the workings of the natural world and the "thought experiment"—an intellectual exercise used by physicists to reach a theoretical conclusion from idealized physical processes. Yet, Einstein was not a purely abstract thinker. He grasped the world in concrete images and strove to translate them into words and equations that could be understood by others.

Einstein's scientific genius lay in his penetrating and intuitive grasp of the workings of the natural world. Yet, his work must be viewed within the context of the evolution of modern physics. He did not create airtight, perfect physical theories that will last forever. He was well aware that his theories were based on a long history of previous efforts of the scientists who had preceded him. Moreover, Einstein declared that just as his theories had gone beyond Newton's, sooner or later someone would have to go beyond his own.

Where the world ceases to be the scene of our personal hopes and wishes, where we face it as free beings admiring, asking, and observing, there we enter the realm of Art and Science. If what is seen and experienced is portrayed in the language of logic, we are engaged in science. If it is communicated through forms whose connections are not accessible to the conscious mind but are recognized intuitively as meaningful, then we are engaged in art. Common to both is the loving devotion to that which transcends personal concerns and volition.
— From the article "What Do the Artistic and Scientific Experiences Have in Common?" Statement published in the German journal *Menschen* upon the request of the editor, Walter Hasenclever, Berlin, January 27, 1921.

OPPOSITE Albert Einstein at the Zurich Polytechnic, 1898
© LOTTE JACOBI COLLECTION, UNIVERSITY OF NEW HAMPSHIRE

The Early Years

Einstein's interest in science and technology was initially stimulated by his father and uncle, who jointly owned an electrotechnical firm. After Einstein completed his secondary education, he moved on to his vocation: the study of physics, especially in its theoretical aspects. His studies at the Zurich Polytechnic (ETH) constituted his scientific apprenticeship. Yet he skipped most of his classes and his knowledge of theory resulted primarily from self-instruction. After his studies he created an informal study group called the "Olympia Academy" with two of his friends.

Upon completion of his studies in 1900, Einstein tried to obtain a position in academia. His efforts met with no success, and in 1902 he finally found employment at the Swiss Patent Office in Bern. Although he was isolated from the scientific community at the time, he kept abreast of significant developments in physics through scientific publications. It was during this period as an unknown patent clerk that he developed and published his first revolutionary theories.

The *annus mirabilis*

For two hundred and fifty years, the basic laws of motion and gravitation as postulated by Newton in the seventeenth century had prevailed. However, by the end of the nineteenth century, serious cracks had developed in the Newtonian edifice. First, Newton had regarded light as a stream of particles, while subsequent experiments showed that light was wave-like. Second, the newly discovered theory of electromagnetism did not fit into the Newtonian

© 1997 BY SIDNEY HARRIS

OPPOSITE Einstein at his desk in the Swiss Patent Office, Bern, 1905

PHOTO BY LUCIEN CHAVAN

system. Various hypotheses—including the ether theory—were put forward, yet none of them was totally satisfactory. Some of these hypotheses foreshadowed Einstein's scientific breakthroughs. However, it was his scientific genius which brought all these elements together and created something entirely new.

6. *Über einen die Erzeugung und Verwandlung des Lichtes betreffenden heuristischen Gesichtspunkt;* von A. Einstein.

[1]

Zwischen den theoretischen Vorstellungen, welche sich die Physiker über die Gase und andere ponderable Körper gebildet haben, und der Maxwellschen Theorie der elektromagnetischen Prozesse im sogenannten leeren Raume besteht ein tiefgreifender formaler Unterschied. Während wir uns nämlich den Zustand eines Körpers durch die Lagen und Geschwindigkeiten einer zwar sehr großen, jedoch endlichen Anzahl von Atomen und Elektronen für vollkommen bestimmt ansehen, bedienen wir uns zur Bestimmung des elektromagnetischen Zustandes eines Raumes kontinuierlicher räumlicher Funktionen, so daß also eine endliche Anzahl von Größen nicht als genügend anzusehen ist zur vollständigen Festlegung des elektromagnetischen Zustandes eines Raumes. Nach der Maxwellschen Theorie ist bei allen rein elektromagnetischen Erscheinungen, also auch beim Licht, die Energie als kontinuierliche Raumfunktion aufzufassen, während die Energie eines ponderabeln Körpers nach der gegenwärtigen Auffassung der Physiker als eine über die Atome und Elektronen erstreckte Summe darzustellen ist. Die Energie eines ponderabeln Körpers kann nicht in beliebig viele, beliebig kleine Teile zerfallen, während sich die Energie eines von einer punktförmigen Lichtquelle ausgesandten Lichtstrahles nach der Maxwellschen Theorie (oder allgemeiner nach jeder Undulationstheorie) des Lichtes auf ein stets wachsendes Volumen sich kontinuierlich verteilt.

Die mit kontinuierlichen Raumfunktionen operierende Undulationstheorie des Lichtes hat sich zur Darstellung der rein optischen Phänomene vortrefflich bewährt und wird wohl nie durch eine andere Theorie ersetzt werden. Es ist jedoch im Auge zu behalten, daß sich die optischen Beobachtungen auf zeitliche Mittelwerte, nicht aber auf Momentanwerte beziehen, und es ist trotz der vollständigen Bestätigung der Theorie der Beugung, Reflexion, Brechung, Dispersion etc. durch das

[2]

[3]

5. *Über die von der molekularkinetischen Theorie der Wärme geforderte Bewegung von in ruhenden Flüssigkeiten suspendierten Teilchen;* von A. Einstein.

In dieser Arbeit soll gezeigt werden, daß nach der molekularkinetischen Theorie der Wärme in Flüssigkeiten suspendierte Körper von mikroskopisch sichtbarer Größe infolge der Molekularbewegung der Wärme Bewegungen von solcher Größe ausführen müssen, daß diese Bewegungen leicht mit dem Mikroskop nachgewiesen werden können. Es ist möglich, daß die hier zu behandelnden Bewegungen mit der sogenannten „Brownschen Molekularbewegung" identisch sind; die mir erreichbaren Angaben über letztere sind jedoch so ungenau, daß ich mir hierüber kein Urteil bilden konnte.

[1]

Wenn sich die hier zu behandelnde Bewegung samt den für sie zu erwartenden Gesetzmäßigkeiten wirklich beobachten läßt, so ist die klassische Thermodynamik schon für mikroskopisch unterscheidbare Räume nicht mehr als genau gültig anzusehen und es ist dann eine exakte Bestimmung der wahren Atomgröße möglich. Erwiese sich umgekehrt die Voraussage dieser Bewegung als unzutreffend, so wäre damit ein schwerwiegendes Argument gegen die molekularkinetische Auffassung der Wärme gegeben.

[2]

§ 1. Über den suspendierten Teilchen zuzuschreibenden osmotischen Druck.

Im Teilvolumen V^* einer Flüssigkeit vom Gesamtvolumen V seien z-Gramm-Moleküle eines Nichtelektrolyten gelöst. Ist das Volumen V^* durch eine für das Lösungsmittel, nicht aber für die gelöste Substanz durchlässige Wand vom reinen Lösungs-

Reprints of Einstein's four revolutionary articles of 1905, his *annus mirabilis*. From left to right:

"On a Heuristic Point of View Concerning the Production and Transformation of Light"

"On the Movement of Small Particles Suspended in Stationary Liquids Required by the Molecular-Kinetic Theory of Heat"

"Does the Inertia of a Body Depend upon Its Energy Content?"

"On the Electrodynamics of Moving Bodies"

In the course of one year—1905—Einstein published four papers which completely revolutionized the concepts of time and space, energy and matter, and resolved some of the fundamental issues which had been baffling physicists for several decades. The year 1905 is known as Einstein's *annus mirabilis,* his miracle year, as 1666 had been for Isaac Newton.

Einstein's first classic paper, which offered an explanation of the photoelectric effect, introduced a radically new understanding of the structure of light. His second paper dealt with Brownian motion and proved the existence of molecules. The two other

13. *Ist die Trägheit eines Körpers von seinem Energieinhalt abhängig?*
von A. Einstein.

Die Resultate einer jüngst in diesen Annalen von mir publizierten elektrodynamischen Untersuchung[1]) führen zu einer sehr interessanten Folgerung, die hier abgeleitet werden soll.

Ich legte dort die Maxwell-Hertzschen Gleichungen für den leeren Raum nebst dem Maxwellschen Ausdruck für die elektromagnetische Energie des Raumes zugrunde und außerdem das Prinzip:

Die Gesetze, nach denen sich die Zustände der physikalischen Systeme ändern, sind unabhängig davon, auf welches von zwei relativ zueinander in gleichförmiger Parallel-Translationsbewegung befindlichen Koordinatensystemen diese Zustandsänderungen bezogen werden (Relativitätsprinzip).

Gestützt auf diese Grundlagen[2]) leitete ich unter anderem das nachfolgende Resultat ab (l. c. § 8):

Ein System von ebenen Lichtwellen besitze, auf das Koordinatensystem (x, y, z) bezogen, die Energie l; die Strahlrichtung (Wellennormale) bilde den Winkel φ mit der x-Achse des Systems. Führt man ein neues, gegen das System (x, y, z) in gleichförmiger Paralleltranslation begriffenes Koordinatensystem (ξ, η, ζ) ein, dessen Ursprung sich mit der Geschwindigkeit v längs der x-Achse bewegt, so besitzt die genannte Lichtmenge — im System (ξ, η, ζ) gemessen — die Energie:

$$l^* = l \frac{1 - \frac{v}{V} \cos \varphi}{\sqrt{1 - \left(\frac{v}{V}\right)^2}}$$

wobei V die Lichtgeschwindigkeit bedeutet. Von diesem Resultat machen wir im folgenden Gebrauch.

[1] 1) A. Einstein, Ann. d. Phys. 17. p. 891. 1905.
2) Das dort benutzte Prinzip der Konstanz der Lichtgeschwindigkeit ist natürlich in den Maxwellschen Gleichungen enthalten.

42*

3. *Zur Elektrodynamik bewegter Körper;*
von A. Einstein.

Daß die Elektrodynamik Maxwells — wie dieselbe gegenwärtig aufgefaßt zu werden pflegt — in ihrer Anwendung auf bewegte Körper zu Asymmetrien führt, welche den Phänomenen nicht anzuhaften scheinen, ist bekannt. Man denke z. B. an die elektrodynamische Wechselwirkung zwischen einem Magneten und einem Leiter. Das beobachtbare Phänomen hängt [1] hier nur ab von der Relativbewegung von Leiter und Magnet, während nach der üblichen Auffassung die beiden Fälle, daß der eine oder der andere dieser Körper der bewegte sei, streng voneinander zu trennen sind. Bewegt sich nämlich der Magnet und ruht der Leiter, so entsteht in der Umgebung des Magneten ein elektrisches Feld von gewissem Energiewerte, welches an den Orten, wo sich Teile des Leiters befinden, einen Strom erzeugt. Ruht aber der Magnet und bewegt sich der Leiter, so entsteht in der Umgebung des Magneten kein elektrisches Feld, dagegen im Leiter eine elektromotorische Kraft, welcher an sich keine Energie entspricht, die aber — Gleichheit der Relativbewegung bei den beiden ins Auge gefaßten Fällen vorausgesetzt — zu elektrischen Strömen von derselben Größe und demselben Verlaufe Veranlassung gibt, wie im ersten Falle die elektrischen Kräfte.

Beispiele ähnlicher Art, sowie die mißlungenen Versuche, eine Bewegung der Erde relativ zum „Lichtmedium" zu kon- [2] statieren, führen zu der Vermutung, daß dem Begriffe der absoluten Ruhe nicht nur in der Mechanik, sondern auch in der Elektrodynamik keine Eigenschaften der Erscheinungen entsprechen, sondern daß vielmehr für alle Koordinatensysteme, für welche die mechanischen Gleichungen gelten, auch die gleichen elektrodynamischen und optischen Gesetze gelten, wie [3] dies für die Größen erster Ordnung bereits erwiesen ist. Wir wollen diese Vermutung (deren Inhalt im folgenden „Prinzip [4] der Relativität" genannt werden wird) zur Voraussetzung erheben und außerdem die mit ihm nur scheinbar unverträgliche

papers Einstein published in 1905 were even more significant for the development of modern physics. They dealt with the nature of space and time and the dynamics of individual moving bodies. Einstein's new theory eventually became known as the special theory of relativity.

The Special Theory of Relativity

Einstein first set out on the trail to the special theory of relativity by thinking about the nature of light. He asked himself what he would see if he could travel alongside a beam of light, at the same speed as the light.

Einstein's special theory of relativity is based on two fundamental postulates. First, in accordance with Newton, the laws of physics are held to be exactly the same for all observers who move at constant velocity relative to one another. Second, adopting Maxwell's electromagnetic theory, Einstein claimed that for such observers, light always moves through empty space at a constant speed, regardless of how the source of light is moving relative to the observer.

Einstein's special theory of relativity constitutes a novel analysis of space and time. It shook the foundations of Newtonian physics. In Newton's universe, both duration and length were absolute and universally the same, no matter what the circumstances.

In Einstein's new relativistic world, the measurement of time and length depends on the relative motion of the observers—especially if the observers are traveling close to the speed of light. A basic assumption of the special theory is the constancy of the speed of light. This assumption leads to all the amazing predictions of the special theory which defy common sense: a ruler moving at a high velocity will seem to shrink and get heavier, while a fast moving clock will seem to run slower.

In 1908, the mathematician Hermann Minkowski provided a geometrization of Einstein's special theory. In addition to the three known dimensions of space, he postulated that time could be regarded as a fourth dimension. Space and time could be regarded as a union, now known as "spacetime." It was Minkowski's geometrization of the special theory which led to widespread acceptance of Einstein's ideas among physicists.

The Photoelectric Effect

In 1900, the German physicist Max Planck had proposed that heat and light radiated from hot objects are emitted or absorbed in discrete chunks of energy he called "quanta." In his paper of 1905, Einstein adopted Planck's theory enthusiastically and used it to explain the photoelectric effect. This effect occurs when a light beam hits a metallic target and causes it to emit electrons. Einstein postulated that a light beam really *does* consist of a stream of quanta (photons), and thus gave the light quantum a physical reality. This seemed to contradict a century of accumulated evidence that light is a form of wave. In fact, light is *both* particle *and* wave. This wave-particle duality lies at the heart of the new quantum physics, developed in the 1920s, and forms the basis of our current understanding of the sub-atomic world. Intriguingly, it was for his explanation of the photoelectric effect, and not for the theory of relativity, that Einstein received the Nobel Prize for Physics. Applications of the photoelectric effect include the television, remote-control devices, automatic door openers, and the CD player.

Brownian Motion

Einstein's second paper of 1905 dealt with
Brownian motion, the agitated dance performed
by microscopic particles suspended in fluids.
Einstein did not set out to explain this phenomenon, but rather worked out, on theoretical grounds
alone, how the particles *ought* to behave. The significance of this paper was that it convinced scientists
that Brownian motion is caused by the motion of
the molecules and that therefore molecules must
really exist.

$E = mc^2$

Einstein's special theory of relativity also led to the
formulation of the most famous scientific equation
of all, $E = mc^2$. This formula, which first appeared
in an article by Einstein in 1907, shows the
equivalence of energy and mass and states that they
are interchangeable: just as some forms of energy
can, under the right circumstances, be turned into
mass, so mass can, under the right circumstances,
be turned into other forms of energy. According to
Einstein's formula, a tiny amount of matter is
equivalent to a vast quantity of energy. However,
utilizing this energy technologically was not
considered feasible in 1905. It was only with the
discovery of nuclear fission in the 1930s that one of
the technological applications of Einstein's formula
became a possibility—the development of nuclear
weapons.

OPPOSITE Manuscript of an article
published in English as "$E = mc^2$: The
Most Urgent Problem of Our Time,"
in *Science Illustrated*, 1946

Das Gesetz von der Äquivalenz von Masse und Energie (E = mc²)

In der vor-relativistischen Physik gab es zwei voneinander unabhängige Erhaltungs bezw. Bilanz gesetze, die strenge Gültigkeit beanspruchten, nämlich

1) den Satz von der Erhaltung der Energie
2) den Satz von der Erhaltung der Masse.

Der Satz von der Erhaltung der Energie, welcher schon von Leibnitz in seiner vollen Allgemeinheit als gültig vermutet wurde, entwickelte sich im 19. Jahrhundert wesentlich als eine Folge eines Satzes der Mechanik. Man betrachte ein Pendel, dessen Masse zwischen den Punkten A und B hin und her schwingt. In A (und B) verschwindet die Geschwindigkeit v, und die Masse liegt um h höher als als im tiefsten Punkte C der Bahn. In C ist diese Hubhöhe verloren gegangen; dafür aber hat die Masse hier eine Geschwindigkeit v. Es ist, wie wenn sich Hubhöhe in Geschwindigkeit und umgekehrt restlos verwandeln könnten. Die exakte Beziehung ist

$$m g h = \frac{m}{2} v^2,$$

wobei g die Beschleunigung der Erdschwere bedeutet. Das Interessante dabei ist, dass diese Beziehung unabhängig ist von der Länge des Pendels und überhaupt von der Form der Bahn in welcher die Masse geführt wird. Interpretation: es gibt ein etwas (nämlich die Energie) das während des Vorgangs erhalten bleibt. In A ist die Energie eine Energie der Lage oder "potentielle Energie" in C eine Energie der Bewegung oder "kinetische Energie". Wenn diese Auffassung das Wesen der Sache richtig erfasst, so muss die Summe

$$m g h + m \frac{v^2}{2}$$

auch für alle Zwischenlagen denselben Wert haben, wenn man mit h die Höhe über C und mit v die Geschwindigkeit in einem beliebigen Punkte der Bahn. Dies verhält sich in der That so. Die Verallgemeinerung dieses Satzes gibt den Satz von der Erhaltung der mechanischen Energie. Wie aber, wenn das Pendel schliesslich durch Reibung zur Ruhe gekommen ist? Davon später.

Beim Studium der Wärme-Leitung war man zu richtigen Ergebnissen gekommen unter Zugrundelegung der Annahme, dass die Wärme ein unzerstörbarer Stoff sei, der vom wärmeren zum kälteren Stoff fliesst. Es schien einen "Satz von der Erhaltung der Wärme" zu geben. Andererseits aber waren seit undenklichen Zeiten Erfahrungen bekannt, nach denen durch Reibung Wärme erzeugt wird (Bewegung der Indianer). Nachdem sich die Physiker lange dagegen

Einstein in his Berlin study, 1916

The General Theory of Relativity

In 1908, Einstein finally acquired a position as a lecturer at the University of Bern. In subsequent years, he rose through the academic hierarchy, taking up posts at the universities of Zurich and Prague and at the Zurich Polytechnic. In 1914, he took up a permanent position as a professor without teaching obligations at the University of Berlin and was appointed a member of the Prussian Academy of Sciences. In 1917, he was appointed director of the Kaiser Wilhelm Institute for Physics.

Einstein began work on a generalization of his theory of relativity six years after the publication of his special theory in 1905. The special theory of relativity only applies in special circumstances—to objects moving at constant speeds in straight lines. A general theory of relativity would include an explanation for objects which move in curved paths and experience acceleration. The most important kind of accelerated motion which makes planets move through space along curved paths and describes the fall of an apple to the ground is caused by gravity.

In 1907, Einstein had what he later described as the "happiest thought of [his] life: for an observer falling freely from the roof of a house there exists— at least in his immediate surroundings—no gravitational field." This insight led Einstein to the postulation of the "equivalence principle," which states that in order for the acceleration caused by gravity to cancel out the force of gravity (leaving the observer in a state of "free fall"), the acceleration and gravity must be exactly equivalent to one another.

Albert Einstein with chemist Fritz Haber in Berlin, July 1, 1914
PHOTO BY SETSURO TAMARU

Einstein lacked the mathematical tools to make
immediate progress and put aside the problem
of gravity for over three years. In 1911, he realized
that the equivalence principle implied an important
and measurable effect: a ray of light passing near
to the sun would be bent. In 1912, his old friend
Marcel Grossmann introduced him to the non-
Euclidian geometry developed by Bernhard
Riemann that was required for the further development of a general theory.

"The Foundation of the General Theory of Relativity." Manuscript of the article published in *Annalen der Physik* (1916).

OPPOSITE Postcard to Pauline Einstein, September 27, 1919. Einstein informs his mother that the general theory of relativity has been confirmed.

Einstein completed the general theory of relativity in 1915 and published it in *Annalen der Physik* in 1916. The article was the first systematic exposé of the general theory of relativity. Einstein donated the manuscript to The Hebrew University on the occasion of its opening in 1925.

Liebe Mutter!

Heute eine freudige Nachricht. H. A. Lorentz hat mir telegraphiert, dass die englischen Expeditionen die Lichtablenkung an der Sonne wirklich bewiesen haben. Maja schreibt mir leider, dass Du nicht nur viel Schmerzen hast, sondern dass Du Dir auch noch trübe Gedanken machest. Wie gern würde ich Dir wieder Gesellschaft leisten, dass Du nicht dem hässlichen Grübeln überlassen wärest! Aber ein Weile werde ich doch hier bleiben müssen und arbeiten. Auch nach Holland werde ich für einige Tage fahren, um mich Ehrenfest dankbar zu erweisen, obwohl der Zeitverlust recht schmerzlich ist.

Dear Mother,

Good news today. H. A. Lorentz has telegraphed me that the British expeditions have definitely confirmed the deflection of light by the sun. Unfortunately, Maja has written me that you're not only in a lot of pain but that you also have gloomy thoughts. How I would like to keep you company again so that you're not left to ugly brooding. But I will have to stay here for a while and work. I will also be traveling to Holland for a few days to show my gratitude to Ehrenfest, even though the delay is rather painful. I truly wish you good days.

Affectionately yours,
Albert

Pauline Einstein
Sanatorium Rosenau
Luzern (Schweiz)

Abs. A. Einstein
Haberlandstr. 5
Berlin.

CARTOON BY CLIFFORD KENNEDY BERRYMAN, DECEMBER 12, 1930; © 1930 THE WASHINGTON POST; REPRINTED WITH PERMISSION

The general theory overturned Newton's theory of gravity, which had been valid for two hundred and fifty years. In Newton's universe, gravity was regarded as a force by which a large mass attracts other masses. The planets were thought to be held in their elliptical orbits around the sun by the force of gravity. In Einstein's universe, gravity is not regarded as an exterior force, but rather as a property of space and time or "spacetime." As gravity affects all matter indiscriminately, the response of a body to gravity is independent of its mass. Gravity is not tied closely to an object itself, but is embedded in the backdrop of space and time— "spacetime" through which an object moves. Space and time are no longer merely the passive arena in which material bodies act out their roles, but are themselves part of the action. Einstein's curved four-dimensional spacetime "continuum" is often likened to a suspended rubber sheet stretched taut but deformed wherever heavy objects—stars, galaxies, or any other matter—are placed on it. Thus, a massive body like the sun curves the space-time around it and the planets move along these curved pathways of spacetime. As Einstein put it: "matter tells space how to bend; space tells matter how to move."

The general theory predicted exactly to what extent a light beam would be bent when it passes near the sun. This prediction was confirmed by observations made by an expedition to West Africa led by the British astronomer Arthur Eddington during a total eclipse of the sun in May 1919. It was the announcement of the confirmation of this predic-tion which thrust celebrity status upon Einstein overnight in November 1919. The general theory continued to be tested and verified throughout the 1920s. Following the development of more sophis-ticated technology in the 1960s, both the special and the general theories of relativity have repeatedly been verified through experiments involving rock-ets, atomic clocks, satellites, and astronauts. One of the predictions of the general theory was that the universe must be expanding, thereby forming the basis for the "big bang" theory. Furthermore, the general theory has been essential in explaining the behavior of bizarre stellar objects within the expanding universe such as black holes and quasars.

Albert Einstein with a group of colleagues in Leiden, early 1920s. *Front row:* Arthur Stanley Eddington, Hendrik Antoon Lorentz. *Back row:* Einstein, Paul Ehrenfest, Willem de Sitter.

Nobel Prize for Physics, 1921: Gold medal with Latin inscription

OPPOSITE Nobel Prize for Physics, 1921: Certificate from the Royal Swedish Academy of Sciences, Stockholm, December 10, 1922

The Nobel Prize

Einstein received the Nobel Prize for Physics for the year 1921. The official announcement was not made until November 1922, while Einstein was on a lecture tour in Japan. The citation states that Einstein received the Nobel Prize "for his services to theoretical physics and in particular for his discovery of the law of the photoelectric effect." Relativity was not mentioned in the citation, mainly because at that time the Swedish Academy considered it too controversial, both scientifically and politically. When Einstein gave his delayed Nobel Prize lecture in Göteborg in July 1923, he ignored the wording of the citation and spoke on the theory of relativity.

Quantum Theory

Following the publication of the general theory of relativity, Einstein resumed his work on sub-atomic processes. In his 1905 paper on the photoelectric effect, Einstein had laid the foundation for the quantum theory. In the period from 1916 to 1925, he made two additional major contributions to quantum theory. First, he confirmed that photons carry momentum. Second, he introduced the notion of stimulated emission of radiation—a concept which eventually led to the development of the laser.

In the mid-1920s, two separate mathematical descriptions of the behavior of electrons were developed by Erwin Schrödinger and Werner Heisenberg. These descriptions formed the basis for a crucial new phase in quantum theory called quantum mechanics. The Danish physicist Niels

KUNGLIGA SVENSKA
VETENSKAPS-AKADEMIEN

har vid sitt sammanträde den 9 November 1922 i enlighet med föreskrifterna i det av

ALFRED NOBEL

den 27 November 1895 upprättade testamente beslutat att, oberoende av det värde som, efter eventuell bekräftelse må tillerkännas relativitets- och gravitationsteorien, överlämna det pris, som för 19 bortgives åt den som inom fysikens område har gjort den viktigaste upptäckt eller uppfinning, till

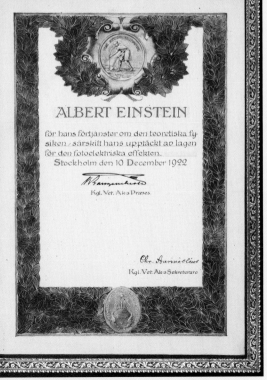

ALBERT EINSTEIN

för hans förtjänster om den teoretiska fysiken, särskilt hans upptäckt av lagen för den fotoelektriska effekten.
Stockholm den 10 December 1922

Kgl. Vet. Aks Præses.

Chr. Aurivillius
Kgl. Vet. Aks Sekreterare.

Albert Einstein and Niels Bohr, 1927

OPPOSITE Einstein in his Berlin study,
late 1920s

Bohr became the major proponent of the dominant interpretation of quantum mechanics known as the "Copenhagen interpretation." The basic assumptions of this interpretation are the following: first, the very act of observing an object changes it. Second, in accordance with the principle of quantum uncertainty, we can never know both the position and the momentum of a particle with absolute precision at the same time. Third, all we *can* ever know is the results of experiments. Quantum mechanics deals *only* with probabilities, not with certainties.

Though he made some of the major contributions to quantum theory, Einstein was vehemently opposed to this new concept of causality. As a realist, he was not prepared to abandon the concept that an objective world exists independently of any subjective observational process.

Einstein's refusal to accept quantum mechanics led to his increased isolation from the scientific mainstream in his later years. Yet he continued to play a crucial role in the development of modern physics: by providing carefully reasoned opposition to quantum mechanics, he forced its adherents to develop stronger justifications for their theories, which were thereby put on a more secure footing. Experiments carried out since Einstein's death have proved that—in spite of its philosophical absurdities, which defy common sense—quantum mechanics does work and can form the theoretical basis for a vast range of technological applications.

EINSTEIN SIMPLIFIED

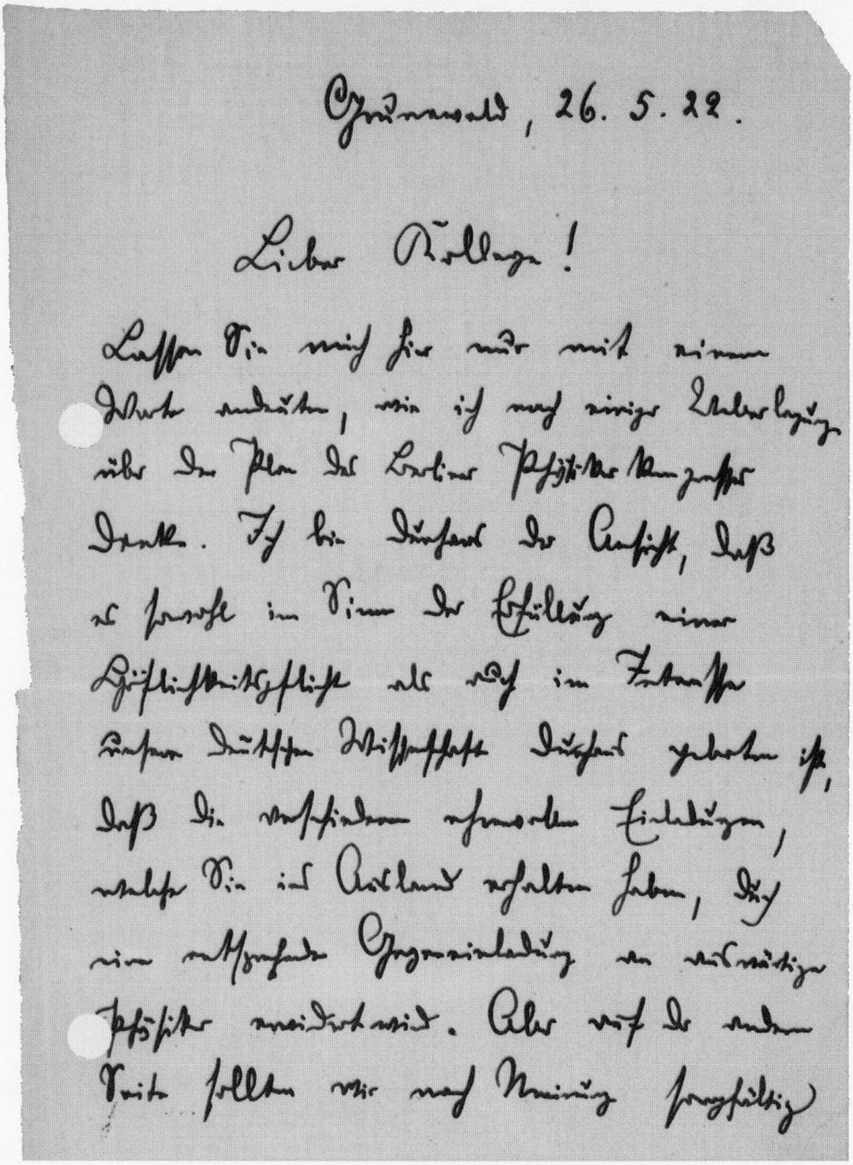

Grunewald, 26. 5. 22.

Lieber Kollege!

Lassen Sie mich hier nur mit einem Worte andeuten, wie ich nach einiger Ueberlegung über den Plan des Berliner Physikerkongresses denke. Ich bin durchaus der Ansicht, dass es sowohl im Sinne der Erfüllung einer Höflichkeitspflicht als auch im Interesse unserer deutschen Wissenschaft durchaus geboten ist, dass die verschiedenen ehrenvollen Einladungen, welche Sie ins Ausland erhalten haben, durch eine entsprechende Gegeneinladung an auswärtige Physiker erwidert wird. Aber auf der andern Seite sollten wir nach Meinung sorgfältig

Letter from Max Planck, May 26, 1922. Planck discusses the reestablishment of ties between German and other European physicists following the First World War.

© DR. HANS ROOS, HILDESHEIM, GERMANY

OPPOSITE Albert Einstein, Paul Ehrenfest, Paul Langevin, Heike Kamerlingh-Onnes, and Pierre Weiss at Ehrenfest's home in Leiden, early 1920s

Zürich, am 4. Dezember 1925.

Hochverehrter Herr Professor!

Haben Sie vielen Dank für Ihren freundlichen Brief vom 14. XI., den ich nur deshalb noch nicht beantwortet habe, weil ich Ihnen sogleich die fertige Ausarbeitung (d. h. _fertig_ von meiner Seite) vorlegen wollte, die ich jetzt mit gleicher Post absende.

Sachlich hat sich gegenüber meiner ersten Mitteilung an Sie nur dies geändert, dass man auch _mit Beibehaltung_ des energielosen Zustandsgebietes die gewöhnlichen Gasgesetze bis zu sehr tiefen Temperaturen herab sehr genau gültig bleiben. Die "Entartungstemperatur" wird nur im Verhältnis e:1 grösser und die Entartung erhält den Charakter einer _Kondensation_, ähnlich wie in Ihrer Theorie der "undulatorischen Moleküle". Gleichwohl lehne ich diese Möglichkeit ab, weil _im vorliegenden Fall_ die letzte Energiestufe _makroskopische_ Grössenordnung bekommt(siehe § 5, Ende).

Ich habe den Autornamen leer gelassen und einige Stellen am Rande rot angestrichen, die rein stilistisch abzuändern wären, wenn Sie mit zeichnen, indem z. B. "ich" durch "der eine von uns" oder durch "wir" zu ersetzen wäre. Daneben werden aber auch Stellen sein, denen Sie vielleicht _sachlich_ nicht ohne weiteres zustimmen, besonders im §1, §6 und §5, Ende, ferner der absolut neutrale Standpunkt, den ich jetzt hinsichtlich der Gewichtszählung (1 oder N!) einnehme. (Nimmt man N! so ist selbstverständlich auch dem abgekühlten Festkörper die Nullpunktsentropie k lg(N!) zuzuschreiben, mit Berufung darauf, dass auch bei der tiefsten Temperatur noch ein, wenn auch noch so kleiner Dampfdruck über dem Körper lagert und dass auf diesem Weg im Lauf der Zeit ein Austausch der Moleküle des Festkörpers sogar wirklich stattfindet.)

Letter from Erwin Schrödinger, December 4, 1925. Schrödinger's correspondence with Einstein shows the influence Einstein had on the development of Schrödinger's wave mechanics.

Letter from Werner Heisenberg, June 10, 1927. Heisenberg's correspondence with Einstein illustrates their fundamental differences of opinion regarding quantum mechanics and causality.

Dear Einstein,

Many thanks for your friendly letter. For all of us, it was a great pleasure to express our feelings on the occasion of your birthday. To speak in the same jocular tone, I cannot help saying about the disquieting questions, that to my mind the issue is not whether we should cling to a reality which is accessible to physical description, but rather, we should pursue the path shown by you and discover the logical prerequisites for the description of the realities. In my impertinent manner, I would even go as far as saying that no one— not even the dear Lord himself— can know what the phrase like playing dice means in this context.

Letter from Niels Bohr, April 11, 1949. Bohr thanks Einstein for his friendly reaction to the *Festschrift* published on the occasion of Einstein's seventieth birthday. In his letter, Bohr alludes to their opposing views on causality. In the 1920s, Einstein had expressed his rejection of quantum mechanics in his famous dictum "God does not play dice," to which Bohr had retorted: "Stop telling God what to do."

© THE NIELS BOHR ARCHIVE, COPENHAGEN

UNIVERSITETETS INSTITUT
FOR
TEORETISK FYSIK

BLEGDAMSVEJ 15 11. April 1949
COPENHAGEN, DENMARK

Lieber Einstein,

Vielen Dank für Ihre freundlichen Zeilen. Es war für uns alle eine grosse Freude, anlässlich Ihres Geburtstages unseren Gefühlen Ausdruck zu geben. Um in demselben scherzhaften Tone zu sprechen, kann ich nicht umhin, über die bangen Fragen zu sagen, dass es sich meines Erachtens nicht darum handelt, ob wir an einer der physikalischen Beschreibung zugänglichen Realität festhalten sollen oder nicht, sondern darum, den von Ihnen gewiesenen Weg weiter zu verfolgen und die logischen Voraussetzungen für die Beschreibung der Realitäten zu erkennen. In meiner frechen Weise möchte ich sogar sagen, dass niemand -und nicht ma der liebe Gott selber- wissen kann, was ein Wort wie wür feln in diesem Zusammenhang heissen soll.

Mit herzlichen Grüssen

Ihr

Niels Bohr

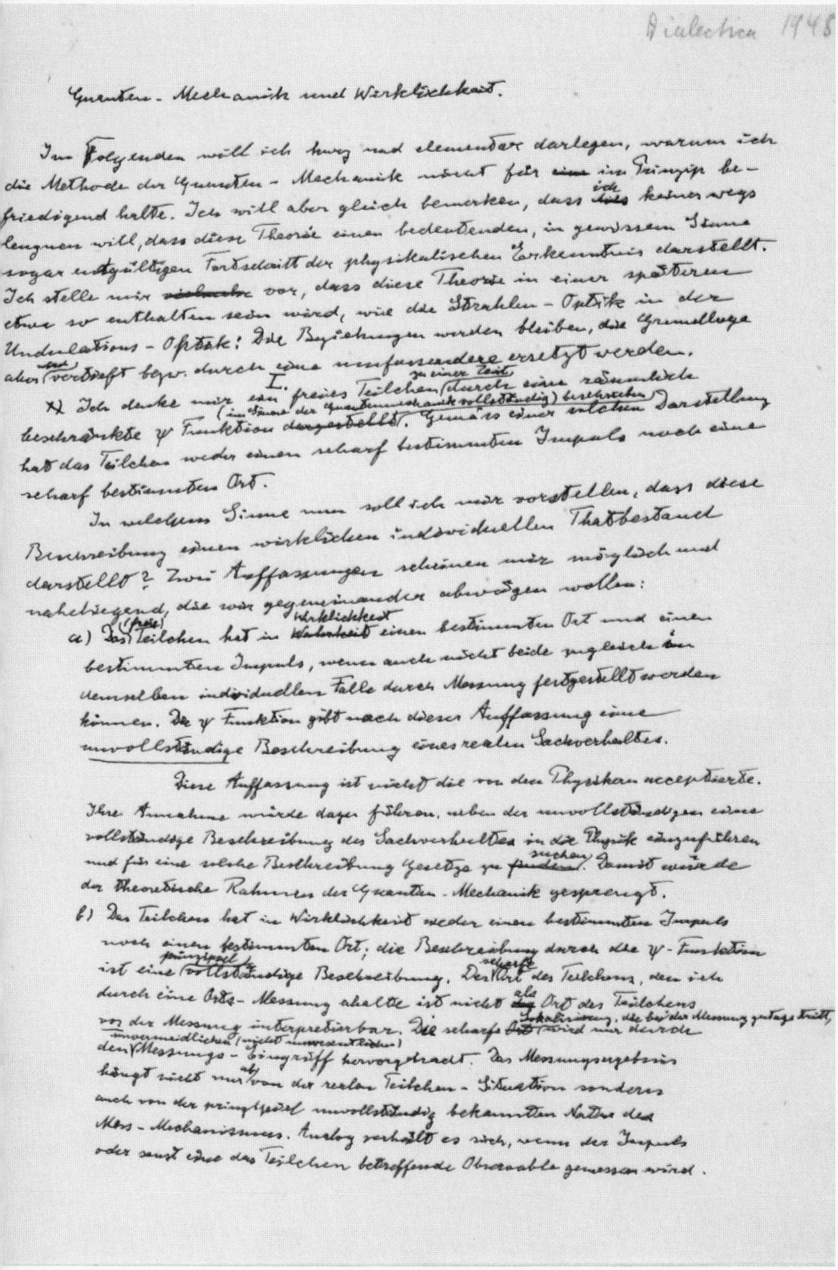

Dialectica 1948

Quanten-Mechanik und Wirklichkeit.

Im Folgenden will ich kurz und elementar darlegen, warum ich die Methode der Quanten-Mechanik nicht für eine im Prinzip befriedigend halte. Ich will aber gleich bemerken, dass ich keineswegs leugnen will, dass diese Theorie einen bedeutenden, in gewissem Sinne sogar endgültigen Fortschritt der physikalischen Erkenntnis darstellt. Ich stelle mir vor, dass diese Theorie in einer späteren etwa so enthalten sein wird, wie die Strahlen-Optik in der Undulations-Optik: Die Beziehungen werden bleiben, die Grundlage aber vertieft bzw. durch eine umfassendere ersetzt werden.

I.

Ich denke mir ein freies Teilchen zu einer Zeit durch eine räumlich beschränkte ψ-Funktion (im Sinne der Quantenmechanik vollständig) beschrieben. Gemäss einer solchen Darstellung hat das Teilchen weder einen scharf bestimmten Impuls noch einen scharf bestimmten Ort.

In welchem Sinne nun soll ich mir vorstellen, dass diese Beschreibung einen wirklichen individuellen Thatbestand darstellt? Zwei Auffassungen scheinen mir möglich und naheliegend, die wir gegeneinander abwägen wollen:

a) Das (freie) Teilchen hat in Wirklichkeit einen bestimmten Ort und einen bestimmten Impuls, wenn auch nicht beide zugleich in demselben individuellen Falle durch Messung festgestellt werden können. Die ψ-Funktion gibt nach dieser Auffassung eine unvollständige Beschreibung eines realen Sachverhaltes.

Diese Auffassung ist nicht die von den Physikern acceptierte. Ihre Annahme würde dazu führen, neben der unvollständigen eine vollständige Beschreibung des Sachverhaltes in die Physik einzuführen und für eine solche Beschreibung Gesetze zu suchen. Damit würde der theoretische Rahmen der Quanten-Mechanik gesprengt.

b) Das Teilchen hat in Wirklichkeit weder einen bestimmten Impuls noch einen bestimmten Ort; die Beschreibung durch die ψ-Funktion ist eine prinzipiell vollständige Beschreibung. Der scharfe Ort des Teilchens, den ich durch eine Orts-Messung erhalte, ist nicht als Ort des Teilchens vor der Messung interpretierbar. Die scharfe Lokalisierung, die bei der Messung zutage tritt, wird nur durch den unvermeidlichen (nicht unwesentlichen) Messungs-Eingriff hervorgebracht. Das Messungsergebnis hängt nicht nur von der realen Teilchen-Situation ab sondern auch von der prinzipiell unvollständig bekannten Natur des Mess-Mechanismus. Analog verhält es sich, wenn der Impuls oder sonst eine das Teilchen betreffende Observable gemessen wird.

In the following, I want to state briefly and elementarily why I believe the method of quantum mechanics is not a satisfactory one in principle. Yet I want to mention right away that I in no way want to deny that this theory represents a significant, in a sense definitive progress of physical knowledge. I can imagine that this theory will be included in a later one in a similar fashion as radiation optics is included in undulation optics. The relations will remain, yet the foundation will be consolidated or replaced by a more comprehensive one.

"Quantum Mechanics and Reality." Manuscript of an article published in *Dialectica* (1948) in which Einstein justifies his opposition to quantum mechanics, yet acknowledges its importance for the development of modern physics.

Unified Field Theory

By the early 1930s Einstein was planning to spend half of his time in Berlin and the rest of his time at the newly founded Institute for Advanced Study at Princeton. When the Nazis came to power, he was in the United States. He decided not to return to Germany but rather to take up permanent residence at Princeton, accepting a lifelong appointment at the Institute in 1933. Although he officially retired in 1945, he continued to work there until shortly before his death in 1955.

During the last thirty years of his life, Einstein's main scientific interest lay in developing a unified field theory, an attempt to explain both gravity and electromagnetism in one broad mathematical structure. He hoped thereby to fill the troubling gap in quantum theory, the inability to describe the world otherwise than in terms of mere probabilities.

Over the years, Einstein proposed unified field theories in various mathematical forms. Einstein himself was usually the first to detect the flaws in them, yet he carried on relentlessly. He published his major attempts in 1923, 1925, 1929, 1931, and 1950.

This quest occupied more of Einstein's years than any other activity. He did not succeed but he was confident that one day someone would. Although it has often been claimed that these were wasted years, it is now believed that Einstein was actually a generation or two ahead of his time.

When Einstein began his quest, the only two forces known to physics were gravity and electromagnetism. Since then, two new nuclear forces have been discovered, the strong and the weak forces.

Einstein with notepad in his Princeton study, 1938

© LOTTE JACOBI COLLECTION, UNIVERSITY OF NEW HAMPSHIRE

OPPOSITE "Unified Field Theory of Gravitation and Electricity." Manuscript of an article published in the Minutes of the Prussian Academy of Sciences, 1925.

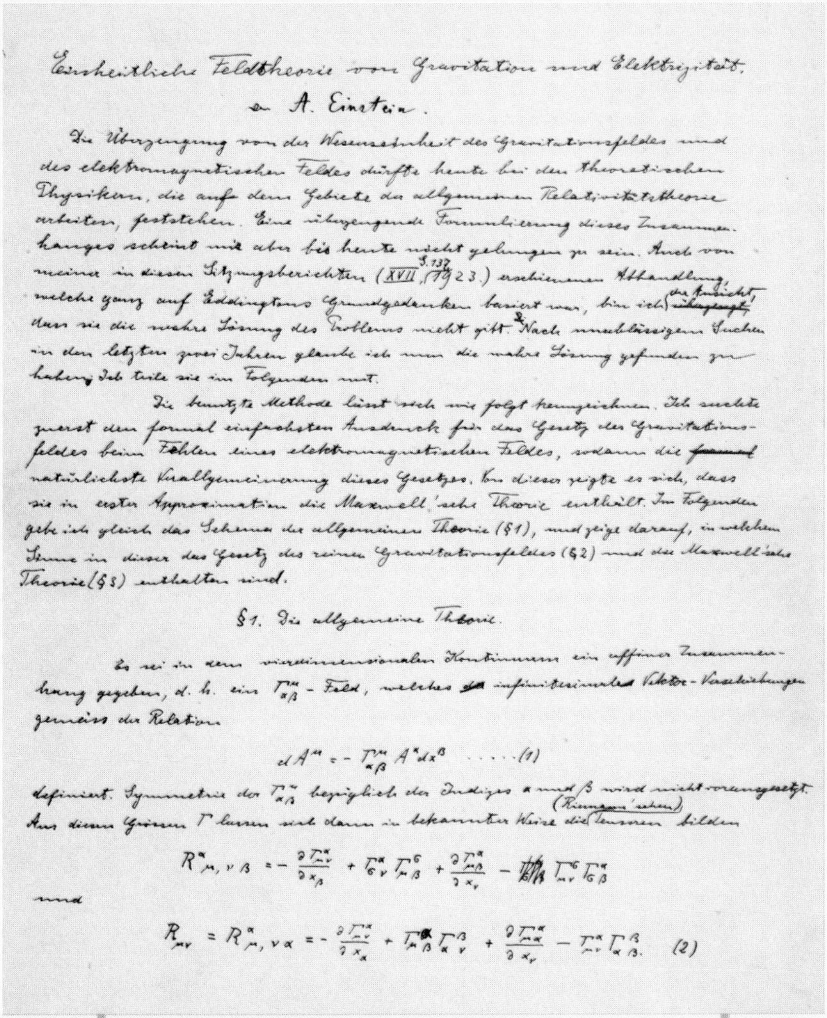

Einheitliche Feldtheorie von Gravitation und Elektrizität.

von A. Einstein.

Die Überzeugung von der Wesenseinheit des Gravitationsfeldes und des elektromagnetischen Feldes dürfte heute bei den theoretischen Physikern, die auf dem Gebiete der allgemeinen Relativitätstheorie arbeiten, feststehen. Eine überzeugende Formulierung dieses Zusammenhanges scheint mir aber bis heute nicht gelungen zu sein. Auch von meiner in diesen Sitzungsberichten (XVII S. 137, 1923) erschienenen Abhandlung, welche ganz auf Eddingtons Grundgedanken basiert war, bin ich der Ansicht, dass sie die wahre Lösung des Problems nicht gibt. Nach unablässigem Suchen in den letzten zwei Jahren glaube ich nun die wahre Lösung gefunden zu haben. Ich teile sie im Folgenden mit.

Die benutzte Methode lässt sich wie folgt kennzeichnen. Ich suchte zuerst den formal einfachsten Ausdruck für das Gesetz des Gravitationsfeldes beim Fehlen eines elektromagnetischen Feldes, sodann die ~~formal~~ natürlichste Verallgemeinerung dieses Gesetzes. Bei dieser zeigte es sich, dass sie in erster Approximation die Maxwell'sche Theorie enthält. Im Folgenden gebe ich gleich das Schema der allgemeinen Theorie (§1), und zeige darauf, in welchem Sinne in dieser das Gesetz des reinen Gravitationsfeldes (§2) und die Maxwell'sche Theorie (§3) enthalten sind.

§1. Die allgemeine Theorie.

Es sei in dem vierdimensionalen Kontinuum ein affiner Zusammenhang gegeben, d. h. ein $\Gamma^{\mu}_{\alpha\beta}$ - Feld, welches infinitesimale Vektor-Verschiebungen gemäss der Relation

$$dA^{\mu} = - \Gamma^{\mu}_{\alpha\beta} A^{\alpha} dx^{\beta} \quad \ldots\ldots (1)$$

definiert. Symmetrie der $\Gamma^{\mu}_{\alpha\beta}$ bezüglich der Indizes α und β wird nicht vorausgesetzt. Aus diesen Grössen Γ lassen sich dann in bekannter Weise die (Riemann'schen) Tensoren bilden

$$R^{\alpha}_{\mu,\nu\beta} = - \frac{\partial \Gamma^{\alpha}_{\mu\nu}}{\partial x_{\beta}} + \Gamma^{\alpha}_{\sigma\nu} \Gamma^{\sigma}_{\mu\beta} + \frac{\partial \Gamma^{\alpha}_{\mu\beta}}{\partial x_{\nu}} - \Gamma^{\sigma}_{\mu\nu} \Gamma^{\alpha}_{\sigma\beta}$$

und

$$R_{\mu\nu} = R^{\alpha}_{\mu,\nu\alpha} = - \frac{\partial \Gamma^{\alpha}_{\mu\nu}}{\partial x_{\alpha}} + \Gamma^{\alpha}_{\mu\beta} \Gamma^{\beta}_{\alpha\nu} + \frac{\partial \Gamma^{\alpha}_{\mu\alpha}}{\partial x_{\nu}} - \Gamma^{\alpha}_{\mu\nu} \Gamma^{\beta}_{\alpha\beta}. \quad (2)$$

Electromagnetism and the strong and the weak forces can be explained by quantum physics—gravity is the odd one out. Physicists are currently trying to develop a quantum theory of gravity as a first stage towards a theory of everything (TOE) which would encompass all known forces and fields of physics in one set of formulas. The ongoing quest for a theory of everything is Einstein's most significant legacy to science.

Einstein participating in Princeton University's bicentennial
procession, 1946

PHOTO BY ALAN WINDSOR RICHARDS, PHOTOGRAPHER FOR PRINCETON UNIVERSITY;
COURTESY BERNICE SHEASLEY

OPPOSITE Einstein in his Princeton study, 1945

PHOTO BY HERMAN LANDSHOFF

Albert Einstein with his assistant Peter G. Bergmann at the
Institute for Advanced Study, October 2, 1940

© LUCIEN AIGNER / CORBIS; COURTESY OF ROBERT KLEIN GALLERY, BOSTON

OPPOSITE Autograph manuscript showing calculations from
the period when Einstein was trying to develop a unified field
theory, Princeton, 1950s

$$m \frac{dx_1}{d\tau} = \text{Impuls in der } x\text{-Axe}$$

$$m \frac{dx_4}{ds} = i \frac{m}{\sqrt{1-u^2}} = i \cdot \text{Energie bis auf add. Konstante}$$

$$m \left(\frac{1}{\sqrt{1-u^2}} - 1 \right) = \text{Kinetische Energie}$$

$$\sum m \frac{dx_i}{d\tau} = \sum \overline{m} \frac{\overline{dx_i}}{d\overline{\tau}} = 0 \quad \text{im speziellen System}$$

$$\text{Energie} = \frac{u}{c_0} + \frac{q}{H} \left(\sum \frac{1}{u} m \frac{dx_4}{ds} \right)$$

$$\sum m \frac{dx_4}{d\tau} \neq \sum \overline{m} \frac{\overline{dx_4}}{d\overline{\tau}} \qquad \text{im speziellen System}$$

Der Vektor $\mathfrak{I}$ besitze im speziellen System nur 4 Komponente. $i\varrho$
Dann ist er in einem andern System $\varrho \, \mathfrak{u}_4$, wobei $\mathfrak{u}_4$ die Geschwind.
Vektor der Transformation ist.

$$\sum m \, u_i' \quad \text{ist solcher Vektor., ebenso } \sum \overline{m} \, \overline{u}_i'$$

~~$\varphi st \; hierbei \; \frac{\sum m}{\sqrt{1-u^2}}$~~

~~Also sind im allgemeinen System dessen Komponenten $\sum m \, u_\nu$~~

Im speziellen System sind die Komponenten dieser beiden Vektore

$$\sum m \quad 0 \quad 0 \quad 0 \quad \frac{i \cdot 2m}{\sqrt{1-u^2}}$$

$$\text{bzw.} \quad 0 \quad 0 \quad 0 \quad \frac{i \cdot 2\overline{m}}{\sqrt{1-\overline{u}^2}}$$

Im allgemeinen System

$$\frac{2m}{\sqrt{1-u^2}} \, v_i \qquad \frac{2m}{\sqrt{1-u^2}} \, v_4$$

$$\text{bzw.} \quad \frac{2\overline{m}}{\sqrt{1-\overline{u}^2}} \, v_i \qquad \frac{2\overline{m}}{\sqrt{1-\overline{u}^2}} \, v_4$$

Damit aber der Impulssatz gelte, muss $\boxed{\dfrac{m}{\sqrt{1-u^2}} = \dfrac{\overline{m}}{\sqrt{1-\overline{u}^2}}}$ sein.

Die Gesamtenergie ist aber $\mathcal{E}_0 + m \left(\frac{1}{\sqrt{1-u^2}} - 1 \right)$

Dieselbe ändert sich nicht durch den Zusammenstoss

Also $\overline{\mathcal{E}}_0 + \overline{m} \left(\frac{1}{\sqrt{1-\overline{u}^2}} - 1 \right)$ ist gleich dem ungestrichenen.

Also $\overline{\mathcal{E}}_0 - \mathcal{E}_0 = \overline{m}_0 - m_0$

Einstein's desk, as he left it after his death in 1955

OPPOSITE Albert Einstein with assistants, Valentine Bargmann
and Peter G. Bergmann, Princeton, October 2, 1940

The Scientist and Society

One of the main incentives for Einstein to embark on a career in science was the independence he believed it offered. Einstein viewed scientific research as a pure quest for knowledge and truth. He was firmly convinced of the inalienable right of the scientist to think, investigate, and act independently. Scientists must have unimpeded access to all the information they require and must be free to discuss work in progress with their colleagues without any restrictions. To his mind, neither the state nor society had the right to interfere with the scientist's freedom of inquiry.

During his lifetime, Einstein perceived a growing threat to the freedom of scientists: first in Europe in the wake of the rise of Nazism and subsequently in the United States as a result of the cold war and McCarthyism. In reaction to these challenges, Einstein maintained that scientists are not merely experts working in a vacuum—they also have some major social responsibilities. They are obliged to anticipate the consequences of their work—if the fruits of their labors are incompatible with their ethical values, they must refuse to participate, even at the risk of loss of employment. Scientists must also work towards increasing society's awareness of the implications of scientific discoveries. In Einstein's opinion, technological applications of scientific theories have the potential to either improve the plight of humankind or to threaten its very existence. It is up to society to decide which option it prefers.

s.Einstein on Peace, p 122, Ch.4"Friede"p 138   16. Febr. 1931
Rede an die Studenten in Pasadena! (Cal, Tech)

Liebe junge Freunde!

Ich freue mich, Sie vor mir zu sehen, eine
blühende Schar junger Menschen, die sich die Technik zum
Lebensberuf erkoren haben.

Ich könnte ein Jubellied singen mit dem
Refrain: Wie herrlich weit haben wirs gebracht, und Ihr werdet
es noch ungeheuer viel weiter bringen. Sind wir doch im Jahrhundert und dazu im Vaterlande der Technik.

Aber es liegt mir ferne so zu sprechen. Vielmehr
geht mir dabei jener Mann ein, der eine nicht sehr anziehende
Frau geheiratet hat und gefragt wird, ob er glücklich sei. Er
antwortete nämlich so: "Wenn ich die Wahrheit sagen wollte,
müsste ich lügen".

So geht es auch mir. Seht Euch einmal einen
echt unzivilisierten Indianer daraufhin an, ob sein Erleben
weniger reich und froh sei als das des zivilisierten Durchschnitts-
menschen! Ich glaube kaum. Es liegt ein tiefer Sinn darin, dass
die Jugend aller zivilisierten Länder mit Vorliebe "Indianer spielt".

Warum beglückt uns so wenig die herrliche, das
Leben erleichternde Arbeit ersparende Technick? Die einfache
Antwort lautet: weil wir noch nicht gelernt haben, einen
vernünftigen Gebrauch von ihr zu machen.

Im Kriege dient sie dazu, dass wir uns gegen-
seitig vergiften oder verstümmeln. Im Frieden hat sie unser
Leben hastig und unsicher gestaltet. Statt uns weitgehend von
geisttötender Arbeit zu befreien, hat sie die Menschen zu
Sklaven der Maschine gemacht, die meist mit Unlust ihr ein-
töniges, langes Tagewerk vollbringen und stets um ihr armseliges
Brot zittern müssen.

Ein hässliches Lied singt uns der alte Mann, werdet
Ihr denken. Ich tue es aber in einer guten Absicht, indem ich
Euch eine Konsequenz nahelegen möchte.

Es genügt nicht, dass Ihr etwas von Technik
verstehet, wenn Eure Arbeit den Menschen einst zum Segen gereichen soll. Die Sorge um die Menschen und ihr Schicksal muss
stets das Hauptinteresse allen technischen Strebens bilden,
die grossen ungelösten Fragen der Organisation der Arbeit
und der Güterverteilung, damit die Erzeugnisse unseres Geistes
dem Menschengeschlecht zum Segen gereichen und nicht zum Fluche.
Vergesst dies nie über Euren Zeichnungen und
Gleichungen.

> Concern for man himself and his
> fate must always form the chief
> interest for all technical endeavors,
> concern for the great unsolved
> problems of the organization of
> labor and the distribution of
> goods—in order that the creations
> of our mind shall be a blessing and
> not a curse for mankind. Never
> forget this in the midst of your
> diagrams and equations.

From a speech held at the California
Institute of Technology, Pasadena,
February 16, 1931

Thus the man of science, as we can observe with our own eyes, suffers a truly tragic fate. In his sincere attempt to achieve clarity and inner independence, he has succeeded, by his sheer super-human efforts, in fashioning the tools which will not only enslave him but also destroy him from within. He cannot escape being muzzled by those who wield political power. He also realizes that mankind can be saved only if a supranational system, based on law, is created to eliminate the methods of brute force. However, the man of science has retrogressed to such an extent that he accepts as inevitable the slavery inflicted upon him by national states. He even degrades himself to such an extent that he obediently lends his talents to help perfect the means destined for the general destruction of mankind. If today's man of science could find the time and the courage to reflect calmly and critically about his plight and the tasks before him, and if he would then act accordingly, the possibilities for a reasonable and satisfactory solution of the present dangerous international situation would be considerably improved.

"On the Moral Obligation of the Scientist." Message sent to the forty-third meeting of the Italian Society for the Progress of Science, October 1950.

The cosmic genius at work

CARTOON BY LOU GRANT

4 | Einstein's Political Activities

"In long intervals I have expressed an opinion on public issues whenever they appeared so bad and unfortunate that silence would have made me feel guilty of complicity."

Einstein the Radical Pacifist

Science was Albert Einstein's first love, yet he always found time to devote tireless efforts to political causes close to his heart. His ardent humanism led him to strive for peace, freedom, and social justice.

The young Einstein found the authoritarianism and militarism of the German educational system profoundly disturbing. The virulent nationalism and brutality of the First World War served to confirm Einstein's pacifist and internationalist convictions. His first overt political act was the signing of an anti-war manifesto in 1914 in which German intellectuals called for a just peace and a "supranational" organization to prevent future wars.

Following the war and the abolition of the German monarchy, Einstein became an unofficial spokesman for the democratic Weimar Republic. In the early 1920s, he took an active role in the Committee for Intellectual Cooperation of the League of Nations, which fostered understanding within the international scientific community. He was a leading member of the German League for Human Rights and spoke out against the spread of fascism and in defense of democracy. Between 1925 and 1932, Einstein became an active leader of the international anti-war movement and supported conscientious objection. In 1932, he invited Sigmund Freud to a public correspondence on how humanity could be spared the continuing menace of war.

The Nazi rise to power brought about a substantial change in Einstein's position: he began to advocate military preparedness by the European democracies against the threat of Nazism.

O that the nations might see, before it is too late, how much of their self-determination they have got to sacrifice in order to avoid the struggle of all against all!
—Statement on nationalism. Excerpt from "Letter to a Friend of Peace" (published in *Mein Weltbild*, 1934).

An exchange of letters between Einstein and Freud
was published by the Committee for Intellectual
Cooperation of the League of Nations in 1933 under
the title *Why War?* The correspondence, reproduced on the following two pages, attempts
an analysis of the causes of war and violence and
explores possible solutions.

Einstein addresses the crowd at a reception
in San Diego, December 31, 1930

The road to international security demands the unconditional renunciation by all nations of part of their freedom of action and sovereignty. I doubt that there is another way to international security. The desire for power makes the ruling party of a nation resist any limitation of its rights to sovereignty. How is it possible that this group, such a small minority, can make subservient to its desires the masses of the people who by a war stand only to lose and to suffer? (In speaking of the masses, I do not exclude soldiers of every rank who have chosen war as their profession, in the belief that they are serving to defend the most precious possessions of their race, and that attack is often the best method of defense.) The immediate answer is: the minority, the ruling class, is in possession of the schools, the church, and the press. By these means it rules and guides the feelings of the majority of the people and bends them to compliance.

Letter to Sigmund Freud, July 30, 1932

Wien im September 1932.

Lieber Herr Einstein!

Als ich hörte, dass Sie die Absicht haben, mich
zum Gedankenaustausch über ein Thema aufzufordern, dem Sie
Ihr Interesse schenken und das Ihnen auch des Interesses
Anderer würdig erscheint, stimmte ich bereitwillig zu. Ich
erwartete, Sie würden ein Problem an der Grenze des heute
Wissbaren wählen, zu dem ein Jeder von uns, der Physiker
wie der Psycholog, sich seinen besonderen Zugang bahnen
könnte, so dass sie sich von verschiedenen Seiten her auf
demselben Boden träfen. Sie haben mich dann durch die
Fragestellung überrascht, was man tun könne,um das Ver-
hängnis des Krieges von den Menschen abzuwehren. Ich
erschrak zunächst unter dem Eindruck meiner - fast hätte
ich gesagt:unserer-Inkompetenz, denn das erschien mir
als eine praktische Aufgabe, die den Staatsmännern zufällt.
Ich verstand dann aber, dass Sie die Frage nicht als Natur-
forscher und Physiker erhoben haben, sondern als Menschen-
freund, der den Anregungen des Völkerbunds gefolgt war,
ähnlich wie der Polarforscher Fridtjof Nansen sich auf sich
genommen hatte, den Hungernden und den heimatlosen Opfern
des Weltkrieges Hilfe zu bringen. Ich besann mich auch,
dass mir nicht zugemutet wird, praktische Vorschläge zu
machen, sondern dass ich nur angeben soll, wie sich das
Problem der Kriegsverhütung e iner psychologischen Be-
trachtung darstellt.

Aber auch hierüber haben Sie in Ihrem Schreiben
das meiste gesagt. Sie haben mir gleichsam den Wind aus
den Segeln genommen, aber ich fahre gern in Ihrem Kielwasser
und bescheide mich damit alles zu bestätigen, was Sie vor-
bringen, indem ich es nach meinem besten Wissen - oder
Vermuten - breiter ausführe.

Now war is in the crassest opposition to the psychical attitude imposed on us by the cultural process, and for that reason we must rebel against it; we simply cannot any longer put up with it. This is not merely an intellectual and emotional repudiation; we pacifists have a constitutional intolerance of war, an idiosyncrasy magnified, as it were, to the highest degree. It seems, indeed, as though the aesthetic humiliation caused by war plays a scarcely smaller part in our rebellion than do its cruelties. And how long shall we have to wait before the rest of mankind become pacifists too? There is no telling. But it may not be utopian to hope that these two factors, the cultural attitude and the justified dread of the consequences of a future war, may result within a measurable time in putting an end to the waging of war. By what paths or by what side-tracks this will come about we cannot guess. But one thing we can say; whatever fosters the growth of culture works at the same time against war.

Letter from Sigmund Freud,
September 1932

Einstein and Rabindranath Tagore, Indian poet and philosopher,
Berlin, 1930. Einstein and Tagore met to discuss contemporary
issues on some three occasions.

LEFT Albert Einstein and George Bernard Shaw in conversation:
Shaw: "Say Einie, do you really think you understand yourself?"
Einstein: "No, Bernie, do you?"

CARTOON BY OLIVER HERFORD; COURTESY OF LIBRARY OF CONGRESS

OPPOSITE "Einstein takes up the sword." Einstein abandons
absolute pacifism.

CARTOON BY CHARLES RAYMOND MACAULEY; COURTESY OF LIBRARY OF CONGRESS;
PUBLISHED IN THE *BROOKLYN EAGLE*, 1933

EINSTEIN'S POLITICAL ACTIVITIES

The Bomb

In popular imagination, Einstein is perceived as the "father of the atomic bomb." Yet this is far from the historical truth.

His famous equation $E = mc^2$ is incorrectly thought to constitute a formula for the construction of nuclear weaponry. Actually, when Einstein devised this equation, he was attempting to provide a theoretical description of physical phenomena and he did not foresee any technological applications. Moreover, the development of the atomic bomb did not depend on Einstein's formula.

It is also widely believed that Einstein had an active role in American efforts to build the bomb. In fact, his only contribution to the development of atomic weapons was the famous letter he sent to President Roosevelt in August 1939. Alarmed by the threat of a Nazi bomb, Einstein appealed to Roosevelt to initiate an American nuclear research program.

Einstein was not involved in the project to develop the atomic bomb known as the Manhattan Project. Indeed, due to his political beliefs he did not receive a security clearance from the U.S. Army. However, he received some information on the progress of the project. He learned of the detonation of the nuclear device over Hiroshima, as did the rest of the world, by public announcement on August 8, 1945.

In the aftermath of the war, Einstein expressed his deep regret for having sent the letter to Roosevelt.

Einstein and the Bomb

CARTOON BY HORACIO GUERRIERO; PUBLISHED IN *URUGUAY TODAY*, 1979

OPPOSITE Einstein reading a manuscript on his Princeton porch, late 1930s

Albert Einstein
Old Grove Rd.
Nassau Point
Peconic, Long Island

August 2nd, 1939

F.D. Roosevelt,
President of the United States,
White House
Washington, D.C.

Sir:

Some recent work by E.Fermi and L. Szilard, which has been communicated to me in manuscript, leads me to expect that the element uranium may be turned into a new and important source of energy in the immediate future. Certain aspects of the situation which has arisen seem
to call for watchfulness and, if necessary, quick action on the part
of the Administration. I believe therefore that it is my duty to bring
to your attention the following facts and recommendations:

In the course of the last four months it has been made probable -
through the work of Joliot in France as well as Fermi and Szilard in
America - that it may become possible to set up a nuclear chain reaction
in a large mass of uranium,by which vast amounts of power and large quantities of new radium-like elements would be generated. Now it appears
almost certain that this could be achieved in the immediate future.

This new phenomenon would also lead to the construction of bombs,
and it is conceivable - though much less certain - that extremely powerful bombs of a new type may thus be constructed. A single bomb of this
type, carried by boat and exploded in a port, might very well destroy
the whole port together with some of the surrounding territory. However,
such bombs might very well prove to be too heavy for transportation by
air.

-2-

The United States has only very poor ores of uranium in moderate
quantities. There is some good ore in Canada and the former Czechoslovakia,
while the most important source of uranium is Belgian Congo.

In view of this situation you may think it desirable to have some
permanent contact maintained between the Administration and the group
of physicists working on chain reactions in America. One possible way
of achieving this might be for you to entrust with this task a person
who has your confidence and who could perhaps serve in an inofficial
capacity. His task might comprise the following:

a) to approach Government Departments, keep them informed of the
further development, and put forward recommendations for Government action,
giving particular attention to the problem of securing a supply of uranium ore for the United States;

b) to speed up the experimental work,which is at present being carried on within the limits of the budgets of University laboratories, by
providing funds, if such funds be required, through his contacts with
private persons who are willing to make contributions for this cause,
and perhaps also by obtaining the co-operation of industrial laboratories
which have the necessary equipment.

I understand that Germany has actually stopped the sale of uranium
from the Czechoslovakian mines which she has taken over. That she should
have taken such early action might perhaps be understood on the ground
that the son of the German Under-Secretary of State, von Weizsäcker, is
attached to the Kaiser-Wilhelm-Institut in Berlin where some of the
American work on uranium is now being repeated.

Yours very truly,

A. Einstein
(Albert Einstein)

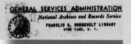

Letter to Franklin D. Roosevelt, U.S. president,
urging him to examine the feasibility of
developing nuclear weapons, August 2, 1939

OPPOSITE Franklin D. Roosevelt's reply to
Einstein's letter, October 19, 1939

THE WHITE HOUSE
WASHINGTON

October 19, 1939

My dear Professor:

I want to thank you for your recent letter and the most interesting and important enclosure.

I found this data of such import that I have convened a Board consisting of the head of the Bureau of Standards and a chosen representative of the Army and Navy to thoroughly investigate the possibilities of your suggestion regarding the element of uranium.

I am glad to say that Dr. Sachs will cooperate and work with this Committee and I feel this is the most practical and effective method of dealing with the subject.

Please accept my sincere thanks.

Very sincerely yours,

Franklin D Roosevelt

Dr. Albert Einstein,
Old Grove Road,
Nassau Point,
Peconic, Long Island,
New York.

May 16, 1954

Mr.H.Herbert Fox
453 North River Str.
Franklin, Ohio.

Dear Sir:

     Thank you for your letter of May 10th. I have
always been a pacifist, i.e. I have declined to
recognize brute force as a means for the solution
of international conflicts. Nevertheless it is,in
my opinion, not reasonable to cling to that principle
unconditionally. An exception has necessarily to be made
if an hostile power threatens wholesale destruction of
one's own group. This was so in the case of Germany
under Hitler. At that time there was the danger that
the Germans would develop atomic weapons which would
have given them the absolute military superiority.
For this reason I had found it necessary that the
American government should be warned of this danger.
If I had known at that time that the danger did,
in fact, not exist I would have done nothing in the
matter.

     Today the danger is a quite different one. There
is no plan of one group to destroy the other. The danger
is now the outbreak of a global war which, in all
probability, would lead to total destruction. Under these
circumstances the ensuring of a peaceful co-existence
of all the nations is not only a moral requirement but
an absolute necessity.

     In view of the present technical means for
destruction no reasonable person can doubt that there
is no other way than the radical pacifist way.

                         Sincerely yours,

                         Albert Einstein.

The Nuclear Threat

With the onset of the atomic era, Einstein realized
that nuclear weapons were a profound risk to
humanity and could bring an end to civilization.
During the last decade of his life, he was tireless
in his efforts to create effective international cooperation to prevent war. He proposed the formation
of a world government to enforce arms control and

May 15,1947

Mr.James B.Burns
225 Henry Hall
Princeton University
Princeton N.J.

Dear Mr.Burns:

    The statement you have heard is,in my opinion, a
slight exaggeration insofar as the radius of the killing
effect of one atomic bomb may be not greater as something
like 10 miles. It seems,therefore,undubitable that one of
those big bombs may be able to destroy the life of a whole
city in one stroke, but not the life in the open country.
This restriction,however, does not alter the fact that
security for the conservation of our national life is
possible only on the basis of powerful supernational
government instead of sovereign national military powers.
It is hard to grasp that there are otherwise reasonable
people able to escape the convincing and irrefutable
arguments which are accessible to everyone.

                    Sincerely yours,

                    Albert Einstein.

establish an international peace-keeping force.
He believed scientists had a special responsibility
to inform their fellow citizens of the perils of
nuclear warfare. In 1946, he became chairman
of the Emergency Committee of Atomic Scientists,
whose goal was to raise funds to support a
massive program of public education and policy
development on atomic energy.

Letter to James B. Burns of Princeton
University, May 15, 1947. In the shadow of
the nuclear threat, Einstein advocates the
establishment of a supranational
government to avert another world war.

OPPOSITE Letter to the American
mathematician H. Herbert Fox, May 18, 1954

In this time of fateful decisions, we must, above all, impress this fact upon our fellow-citizens: whenever the belief in the omnipotence of physical force dominates the political life of a nation, this force takes on a life of its own and becomes even stronger than the very men who intended to use it as a tool. The proposed militarization of the nation not only immediately threatens us with war; it will also slowly but surely undermine the democratic spirit and the dignity of the individual in our land. The assertion that events abroad are forcing us to arm is incorrect; we must combat this false assumption with all our strength. Actually, our own rearmament, because of its effect upon other nations, will bring about the very state of affairs upon which the advocates of armaments seek to base their proposals.
—Statement of acceptance of the "One World Award," Carnegie Hall, New York, April 28, 1948. Einstein warns of the dangers of militarization and the arms race.

Einstein at his home in Princeton, New Jersey, with Elliot Roosevelt. On this occasion, Einstein was interviewed for television by Eleanor Roosevelt. In their exchange, she asked Einstein for his opinion on the construction of the hydrogen bomb, February 10, 1950.

PHOTOS COURTESY OF FRANKLIN D. ROOSEVELT LIBRARY

EINSTEIN'S AXIOM

Dear Mr. Frauenglass:

Thank you for your communication. By "remote field" I referred to the theoretical foundations of physics.

The problem with which the intellectuals of this country are confronted is very serious. The reactionary politicians have managed to instill suspicion of all intellectual efforts into the public by dangling before their eyes a danger from without. Having succeeded so far they are now proceeding to suppress the freedom of teaching and to deprive of their positions all those who do not prove submissive, i. e., to starve them.

What ought the minority of intellectuals to do against this evil? Frankly, I can see only the revolutionary way of non-cooperation in the sense of Gandhi's. Every intellectual who is called before one of the committees ought to refuse to testify, i. e., he must be prepared for jail and economic ruin, in short, for the sacrifice of his personal welfare in the interest of the cultural welfare of his country.

This refusal to testify must be based on the assertion that it is shameful for a blameless citizen to submit to such an inquisition and that this kind of inquisition violates the spirit of the Constitution.

If enough people are ready to take this grave step they will be successful. If not, then the intellectuals of this country deserve nothing better than the slavery which is intended for them.

Sincerely yours,
A. Einstein.

P. S. This letter need not be considered "confidential."

In 1953, William Frauenglass, an American teacher, was summoned to testify before the House Un-American Activities Committee and wrote to Einstein, asking for his advice. In his reply, Einstein advocates Gandhi's policy of non-violence as the method to combat McCarthyism; published in the *New York Times*, July 12, 1953.

Einstein criticized both the United States and the USSR for their cold war policies of mutual fear and suspicion and endeavored to bring about cooperation between scientists of the two countries.

One of his last political acts was his signing of the Russell-Einstein manifesto—jointly with the philosopher Bertrand Russell. The declaration constitutes a solemn warning concerning the appalling consequences of a nuclear war and an appeal for nuclear disarmament. It formed the basis for the subsequent Pugwash Conferences, an influential forum of scientists held for the discussion of the nuclear threat.

Concern for Civil Liberties

After the Second World War, Einstein was gravely concerned with the spread of a cold war mentality in the United States and Senator Joseph McCarthy's hounding of leftists and liberals.

He was looked to as a sympathetic, moral figure by those summoned to appear before the House Un-American Activities Committee. Asked for his advice on how they should defend themselves, Einstein publicly advised the accused to follow Mahatma Gandhi's principle of civil disobedience and to refuse to cooperate with the Committee. Twenty-five years after his death, it was revealed that Einstein himself had been under investigation by the U.S. authorities. The FBI disclosed that it had kept a massive file on his political activities from 1932 until his death.

Einstein was also concerned with the plight of ethnic minorities in the United States and supported the fledgling black civil rights movement.

Einstein meeting with Henry Wallace, a left-wing U.S. politician on the election trail. Also pictured are Frank Kingdon, a columnist, and Paul Robeson, the baritone and civil rights campaigner.

© BETTMANN/CORBIS

Einstein receiving an honorary degree from Lincoln University, Pennsylvania (originally an all-male institution founded to educate African Americans), May 3, 1946

© PEACE PHOTO

In the early 1930s, not everyone welcomed Einstein to the United States. A contemporary cartoon depicts Einstein and the Daughters of the American Revolution, an organization of patriotic women, protesting against the U.S. authorities granting Einstein an entry visa. The caption reads: "You go back where you came from Mr. Einstein. We got plenty of unemployed astrologers of our own."

October 13,1954

Editor,The Reporter
220 East 42nd Str.
New York City

Dear Sir:

        You have asked me what I thought about
your articles concerning the situation of the
scientists in America. Instead of trying to analyze
the problem I may express my feeling in a short remark:
If I would be a young man again and had to decide how
to make my living, I would not try to become a scientist
or scholar or teacher. I would rather chose to be a
plumber or a peddlar in the hope to find that modest
degree of independence still available under present
circumstances.

                Sincerely yours,

                Albert Einstein.

In 1954, the editor of *The Reporter* wrote to Einstein to ask for a statement regarding the situation of intellectuals under McCarthy. Einstein replied with his famous "plumber statement," October 13, 1954.

OPPOSITE Letter from the Stanley Plumbing and Heating Co., November 11, 1954. One of the many humorous reactions to Einstein's "plumber statement."

STANLEY PLUMBING & HEATING CO.
CONTRACTORS
1212 ~~1851~~ SIXTH AVENUE
NEW YORK 19, N. Y.
—
JU 2-4846

November 11, 1954

Dr. Albert Einstein
Princeton University
Princeton, New Jersey

Dear Dr. Einstein:

  As a plumber, I am very much interested in
your comment made in the letter being published in the
Reporter Magazine. Since my ambition has always been
to be a scholar and yours seems to be a plumber, I
suggest that as a team we would be tremendously successful.
We can then be possessed of both knowledge and independence.

  I am ready to change the name of my firm to read:
Einstein and Stanley Plumbing Co.

    Respectfully yours,

    R. Stanley Murray

RSM:hks

Einstein and his secretary, Helen Dukas, in his Princeton study, October 2, 1940

OPPOSITE "Why Socialism?" Article published in *Monthly Review* (May 1949). Einstein expresses his criticism of capitalism and Soviet communism and his support for ethical socialism.

Social and Economic Views

A pacifist in political affairs, Einstein was a committed socialist on social and economic issues. To him the purpose of socialism was to overcome the predatory phase of human development characterized by conquest, subjugation, and exploitation. In his opinion, a capitalist economy renders society vulnerable to severe cycles of economic depression with devastating social consequences. He supported a moderately planned economy to control the excesses of both capitalism and Soviet communism.

He believed that an egalitarian education system would foster the political and social values necessary to create just societies.

Warum Sozialismus?

[Handwritten manuscript in German — Albert Einstein's draft of "Why Socialism?"]

This crippling of individuals I consider the worst evil of "capitalism." Our whole educational system suffers from this evil. An exaggerated competitive attitude is inculcated into the young individual, who is trained to worship acquisitive success as a preparation for his future career. I am convinced there is only one way to eliminate these grave evils, namely through the establishment of a socialist economy, accompanied by an educational system which would be oriented towards social goals. In such an economy, the means of production are owned by society itself and are utilized in a planned fashion. Nevertheless, a planned economy is not yet socialism. A planned economy as such may be accompanied by the complete enslavement of the individual. The achievement of socialism requires the solution of some extremely difficult socio-political problems: how is it possible, in view of the far-reaching centralization of political and economic power, to prevent bureaucracy from becoming all-powerful and overweening? How can the rights of the individual be protected and herewith a democratic counterweight to the power of bureaucracy be assured?

וועידת הסטודנטים העברים בגרמניה
אידישע סטודענטן-קאנפערענץ אין דײטשלאנד
JÜDISCHE STUDENTENKONFERENZ IN DEUTSCHLAND

the most elevated spiritual and intellectual tradition
the Jewish people has established through its best min
hearts both in antiquity and in modern times. Our firs

5 | Einstein's Jewish Identity

"The pursuit of knowledge for its own sake, an almost fanatical love of justice, and the desire for personal independence—these are the features of the Jewish tradition which make me thank the stars that I belong to it."

Einstein with the famous Yiddish actor
Maurice Schwartz and his troupe after a
performance of *Yoshe Kalb*, Princeton, 1934

OPPOSITE "Christianity and Judaism,"
statement for the Rumanian Jewish journal
Renasterea Noastra, January 1933

Einstein the Jew

Throughout his life, Albert Einstein felt a close
affinity with the Jewish people. He was born into
an assimilated German-Jewish family whose milieu
fostered in him a strong Jewish identity. Einstein
defined Judaism as a culture with a shared historical past and common ethical values rather than as
an institutionalized religion. For him the main values of Judaism were intellectual aspiration and the
pursuit of social justice. Like Spinoza, he did not
believe in a personal god, but that the divine reveals
itself in the physical world.

If one purges the Judaism of the Prophets and Christianity as Jesus Christ taught it of all subsequent additions, especially those of the priests, one is left with a teaching which is capable of curing all the social ills of humanity. It is the duty of every man of good will to strive steadfastly in his own little world to make this teaching of pure humanity a living force, so far as he can. If he makes an honest attempt in this direction without being crushed and trampled underfoot by his contemporaries, he may consider himself and the community to which he belongs lucky.

It seems to me that the idea of a personal God is an anthropomorphic concept which I cannot take seriously. I feel also not able to imagine some will or goal outside the human sphere. My views are near to those of Spinoza: admiration for the beauty of and belief in the logical simplicity of the order and harmony which we can grasp humbly and only imperfectly. I believe that we have to content ourselves with our imperfect knowledge and understanding and treat values and moral obligations as a purely human problem—the most important of all human problems.
—Einstein expresses his views on God and religion. Letter to Mr. Murray Gross of Brooklyn, New York, requesting Einstein to expound upon his belief in God in light of his insight into the mechanics of nature, April 25, 1947.

> In the last analysis, everyone is a human being, irrespective of whether he is an American or a German, a Jew or a Gentile. If it were possible to manage with this point of view, which is the only dignified one, I would be a happy man. I find it very sad that divisions according to citizenship and cultural tradition should play so great a role in modern practical life. But since this cannot be changed, one should not close one's eyes to reality.

Princeton N.J., den 3.April 1935

Herrn Gerald M.Donahue
9105 Colonial Road
Brooklyn-New York

Sehr geehrter Herr:

Im letzten Grunde ist jeder ein Mensch, gleichgültig ob Amerikaner, Deutscher, Jew or Gentile. Wenn es möglich wäre, mit diesem allein würdigen Standpunkt auszukommen, wäre ich ein glücklicher Mensch. Ich finde es traurig, dass im heutigen praktischen Leben Trennungen nach Staatszugehörigkeit und kultureller Tradition eine so erhebliche Rolle spielen. Da dies nun aber einmal unabänderlich ist, darf man sich der Wirklichkeit gegenüber nicht verschliessen.

Was nun die eine alte Traditionsgemeinschaft bildende Judenheit anbelangt, so lehrt deren Leidensgeschichte, dass - mit den Augen des Historikers gesehn - das Jude-Sein sich in stärkerem Masse ausgewirkt hat als die Zugehörigkeit zu politischen Gemeinschaften. Wenn zum Beispiel die deutschen Juden aus Deutschland vertrieben werden, so hören sie auf, Deutsche zu sein, ändern ihre Sprache und ihre politische Zugehörigkeit, aber sie bleiben Juden. Warum dies so ist, ist gewiss eine schwierige Frage; ich sehe den Grund in der Hauptsache nicht in Merkmalen der Rasse, sondern in fest eingewurzelten Traditionen, die sich keineswegs auf das Religiöse beschränken. An dieser Tatsache wird dadurch nichts geändert, dass Juden als Bürger bestimmter Staaten in deren Kriegen zum Opfer fallen.

Mit ausgezeichneter Hochachtung

Einstein expresses his views on ethnic and cultural diversity and national conflicts. Letter to Gerald M. Donahue, an American who wrote to Einstein expressing his view that Jews were first and foremost citizens of their countries, April 3, 1935.

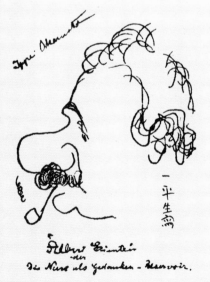

LEFT "The nose as a reservoir for thoughts"
CARTOON BY IPPEI OKAMOTO, 1922; COURTESY AIP EMILIO SEGRÈ VISUAL ARCHIVES

The bond that has united the Jews for thousands of years and that unites them today is, above all, the democratic ideal of social justice, coupled with the ideal of mutual aid and tolerance among all men. Even the most ancient religious scriptures of the Jews are steeped in these social ideals, which have powerfully affected Christianity and Mohammedanism and have had a benign influence upon the social structure of a great part of mankind. Personalities such as Moses, Spinoza, and Karl Marx, dissimilar as they may be, all lived and sacrificed themselves for the ideal of social justice; and it was the tradition of their forefathers that led them on this thorny path. The second characteristic trait of Jewish tradition is the high regard in which it holds every form of intellectual aspiration and spiritual effort. I am convinced that this great respect for intellectual striving is solely responsible for the contributions that the Jews have made towards the progress of knowledge, in the broadest sense of the term.

Einstein defines the social characteristics of the Jewish people and affirms his belief in humanistic Judaism. From "Why Do They Hate the Jews?" published in *Collier's Weekly*, November 26, 1938.

Jewish Nationalism

Prior to 1914, Einstein did not display an overt interest in Jewish affairs. However, his encounter with German anti-Semitism, directed against East European Jewish immigrants during and after the First World War, made him reassess the place of Jews in German society. Like many other German-Jewish intellectuals he came to reject Jewish attempts at total assimilation and to believe that Jewish self-confidence could be restored through a cultural renaissance. Despite his distaste for every form of nationalism, he came to adopt a cultural and non-political definition of Jewish nationalism akin to that of Ahad Ha'am. Therefore, the cultural aspects of German Zionism greatly appealed to him.

For Einstein the aim of establishing a Jewish home-land in Palestine should be the creation of a spiritual center, a model society. He supported colonization efforts in Palestine to provide a refuge for Eastern European Jews, but saw the settlement of Palestine mainly as an instrument for enhancing the cultural identity and social cohesion of Western Jewry.

Einstein did not see any moral justification or practical possibility for carrying out Zionism in the absence of cooperation with the Arab population. Therefore, he urged for a solution of the Arab-Jewish conflict in Palestine based on mutual understanding and consent. Until the summer of 1947, he advocated a binational solution in Palestine. In the light of the war of 1948, he resigned himself to a solution involving partition. Following independence, he strongly supported the State of Israel, yet remained highly critical

Albert Einstein and Chaim Weizmann during their visit to New York, April 1921

Einstein visits the British High Commissioner's residence at Augusta Victoria Hospital, Jerusalem, February 1923. Front row includes, left to right: Elsa Einstein, Herbert Samuel, Lady Beatrice Samuel, Einstein, and Father Dhomme. Back row includes, left to right: Father Sertillange, Norman Bentwich, Mrs. Richmond, Ernest T. Richmond, and Father G. Orfali.

PHOTO BY FATHER CARRIÈRE

Reception for Einstein at the Municipality of Tel Aviv at which he was made an honorary citizen of the city, February 8, 1923. Front row includes Elsa and Albert Einstein with Mayor Meir Dizengoff and A. L. Esterman. Second row includes Avraham Mibashan, Yehuda Grazovsky, Bezalel Jaffe, Ahad Ha'am, Yosef Zeidner, Theodor Zlocisti, David Ijmojic, Shmuel Tulkovsky, and Benzion Mossinson.

PHOTO COURTESY OF THE CENTRAL ZIONIST ARCHIVES, JERUSALEM

of its political leadership. He warned of the pitfalls of narrow nationalism inherent in statehood. In his view, Israel's treatment of its Arab minority would be the test of its moral integrity.

Palestine is for us Jews not a mere matter of charity or colonization; it is a problem of paramount importance for the Jewish people. Palestine is first and foremost not a refuge for east European Jews, but the incarnation of a reawakening sense of national solidarity for all Jews. But is it opportune to revive and to strengthen this sense of solidarity? To that question I must reply with an unqualified affirmative, not only because that answer expresses my instinctive feeling, but also, I believe, on rational grounds. Nationalities do not want to be fused; they each want to go their own way. A satisfactory situation can be brought about only if they mutually tolerate and respect one another. This demands above all things that we Jews become once more conscious of our existence as a nation and regain the self-respect which we require for a prosperous existence. We must learn once more to enthusiastically declare our loyalty to our ancestry and our history; we must once more take upon ourselves, as a nation, cultured tasks of a kind calculated to strengthen our feeling of solidarity. It is not sufficient for us to participate as individuals in the cultural development of mankind: we must also set our hands to such tasks which can only be accomplished by national communities. In this way and in this way only can the Jewish people regain its social health.

Einstein explains the importance of the spiritual and cultural renaissance of the Jewish people. From a speech to U.S. Zionists, New York, April 1921.

ABOUT ZIONISM

Speeches and Letters

by

PROFESSOR ALBERT EINSTEIN

Translated and Edited

With an Introduction by

LEON SIMON

London
THE SONCINO PRESS
5 Gower Street
1930

Albert Einstein: *About Zionism: Speeches and Letters*, ed. Leon Simon (London, 1930)

THE JEWS AND PALESTINE

(1)

The Palestine problem, as I see it, is twofold. There is first the business of settling the Jews in the country. This demands external assistance on a large scale; it cannot be successfully accomplished unless the national resources of Jewry are laid under contribution. The second task is that of stimulating private initiative, especially in the commercial and industrial spheres.

The deepest impression left on me by Zionist work in Palestine is that of the self-sacrifice of the young men and women workers. Gathered here from all sorts of different environments, they have succeeded, under the influence of a common ideal, in forming themselves into closely-knit communities and in working together on lines of systematic co-operation. I was also most favourably impressed by the spirit of initiative shown in the urban development. There is something here that almost suggests an avalanche. One feels that the work is being borne along on the wings of a strong national sentiment. Nothing else could explain the extraordinarily rapid advance, especially on the sea-coast near Tel-Aviv.

41

Went with Sir Herbert Samuel
on foot into the city (Sabbath!),
walked on path past city walls to
picturesque, old gate. Path into
the city bathed in sunshine. Hard,
barren hilly landscape with white,
stone houses, mostly dome-topped,
and blue sky, breathtakingly
beautiful, as is the city squeezed
into the square walls. Continued
into the city with Ginzberg.*
Through the bazaar and other
narrow alleys to the large mosque
on a splendidly wide, raised square
where Solomon's temple stood.
Resembles Byzantine church,
polygonal, with a central dome
supported by columns. On the
other side of the square a basilica-
like mosque of mediocre taste.
Then down to the temple wall
(Wailing Wall) where dull-minded
fellow Jews prayed out loud, their
faces turned to the wall, bending
their bodies backwards and
forwards in a rocking movement.
Pitiful sight of people with a
past but no present. Then zig-zag
through the (very dirty) city,
swarming with all kinds of holy
men and members of various races,
noisy and strangely oriental.

*Shlomo Ginzberg-Ginossar, son
of Ahad Ha'am, accompanied
Einstein throughout his visit to
Palestine.

Einstein's impressions of Jerusalem during his only visit to Palestine.
From his travel diary, February 3, 1923.

OPPOSITE, ABOVE Einstein addresses the Conference of Jewish
Students in Germany, February 27, 1924
PHOTO DONATED BY EFRAT KARMON, JERUSALEM

OPPOSITE, BELOW Einstein on the Jewish "inferiority complex."
Letter to Martin Rotke, a young Jew who wrote to Einstein to ask for
his advice on how to act in the face of the first revelations of the
atrocities of the Holocaust, October 5, 1943.

October 5th, 1943

Mr. Martin Rotke
553 Second Ave.
San Francisco, 18, Cal.

Dear Sir:

     I received your letter of September 21rst.
The relationship of a Jew to his environment involves
difficult problems. For us our outside situation is
a given fact not depending very much from our behavior.
The most important thing is - in my opinion - that we
aquire enough self-reliance so that we lose the dependence
and the hyper-sensitiveness towards the non-Jewish world.
We have to aquire a cool understanding without bitterness
of the prejudices and the corresponding behavior towards
us, so that no feeling of inferiority can develop in
ourself. This is not only true in the Jewish sphere but
everywhere where passion and prejudice impair human
relationships. Such burden may develop independence and
character of stronger individuals, but unfortunately also
injure weak personalities. Spinoza is an admirable example
of the first kind. Sound humor will also help.

     Sincerely yours,

     Professor Albert Einstein.

One who, like myself, has cherished for many years the conviction that the humanity of the future must be built up on an intimate community of the nations, and that aggressive nationalism must be conquered, can see a future for Palestine only on the basis of peaceful co-operation between the two peoples who are at home in the country. For this reason I should have expected that the great Arab people will show a truer appreciation of the need which the Jews feel to re-build their national home in the ancient seat of Judaism; I should have expected that by common effort ways and means would be found to render possible an extensive Jewish settlement in the country. I am convinced that the devotion of the Jewish people to Palestine will benefit all the inhabitants of the country, not only materially, but also culturally and nationally. I believe that the Arab renaissance in the vast expanse of territory now occupied by the Arabs stands only to gain from Jewish sympathy. I should welcome the creation of an opportunity for absolutely free and frank discussion of these possibilities, for I believe that the two great Semitic peoples, each of which has in its way contributed something of lasting value to the civilisation of the West, may have a great future in common, and that instead of facing each other with barren enmity and mutual distrust, they should support each other's national and cultural endeavours, and should seek the possibility of sympathetic co-operation. I think that those who are not actively engaged in politics should above all contribute to the creation of this atmosphere of confidence.

I deplore the tragic events of last August not only because they revealed human nature in its lowest aspects, but also because they have estranged the two peoples and have made it temporarily more difficult for them to approach one another. But come together they must, in spite of all.

Letter to Azmi El-Nashashibi, editor of the Palestinian Arab newspaper *Falastin*, January 28, 1930

PORT EINSTEIN
BERLIN W., den 25. November 29.
HAMBERLANDSTR. 5

Herrn Professor Dr. Weizmann
London W.14
Oakwood, 16, Addison Crescent

Lieber Herr Weizmann!

Ich danke Ihnen bestens für Ihren Brief und kann mir denken, dass Sie von schweren Sorgen erfüllt sind. Gleichzeitig aber muss ich Ihnen offen sagen, dass mich die Haltung unserer leitenden Männer beunruhigt. Neulich hat Brodetzki in einem Agency-Vortrag mit jener Aeusserlichkeit und Oberflächlichkeit wieder das arabische Problem behandelt, die den gegenwärtigen Zustand der Dinge herbeigeführt hat. Das wirtschaftliche und psychologische Problem der Jüdisch-arabischen Symbiose wurde überhaupt nicht berührt, sondern der Konflikt als Episode behandelt. Dies war umso unangebrachter, als die vernünftigeren Zuhörer von der Unaufrichtigkeit einer solchen Betrachtungsweise voll überzeugt sein werden. Ich sende Ihnen hier einen Brief von Hugo Bergmann, der nach meiner Ueberzeugung das Wesentliche trifft. Wenn wir den Weg ehrlicher Kooperation und ehrlichen Paktierens mit den Arabern nicht finden werden, so haben wir auf unserem zweitausendjährigen Leidensweg nichts gelernt und verdienen das Schicksal, das uns treffen wird. Insbesondere müssen wir uns nach meiner Meinung davor hüten, uns zuviel

Einstein advocates reconciliation with the Arab population of Palestine. Letter to Chaim Weizmann, November 25, 1929.

If we do not succeed in finding the path of honest cooperation and coming to terms with the Arabs, we will not have learned anything from our two-thousand-year-old ordeal and will deserve the fate which will beset us.

THE INSTITUTE FOR ADVANCED STUDY
SCHOOL OF MATHEMATICS
FINE HALL
PRINCETON, NEW JERSEY

April 18th 1938

Dr.Kathryn McHall
Director
American Association of University Women
1634 Eye Str. N.W.
Washington D.C.

Dear Mrs.McHall:

I am writing you in the interest of Dr.Marietta Blau
who has worked for some years at the Institute for Radium-
forschung in Vienna. She is of Jewish race and has therefore
to leave Austria resp.Germany.

I am interested in Miss Blau as she has done outstanding
original scientific work and is estimated very highly by her
colleagues. Her field of research is nuclear physics and especially
investigation of cosmic ray particles using a simple method which
she has worked out. You would oblige me in telling me how it
may be possible to find a position for Miss Blau where she can
continue her research. I suppose that she could also fill a
teaching position but I think that she should find some opportunity
for doing research.

Very sincerely yours,

Professor Albert Einstein.

One of the numerous cases in which
Einstein intervened on behalf of Jewish
refugees from Germany and Austria. Letter
to Dr. Kathryn McHall, American Association
of University Women, April 18, 1938.

THIS ITEM WAS GENEROUSLY DONATED BY THE AAUW.

The Holocaust

Einstein, himself an émigré from Nazi Germany to
the United States, devoted himself to finding new
homes and jobs for countless Jewish and political
refugees. He was particularly concerned with the
plight of Jewish intellectuals from Germany and
Austria and supported the establishment of special
universities to accommodate them. When the full
dimension of the Holocaust became known,
Einstein was deeply affected. Shaken by the extent
of man's inhumanity to man, he was never to rec-
oncile himself with Germany again.

Die feierliche ~~Zusammenkunft~~ Versammlung des heutigen Tages hat eine tiefe Bedeutung. Wenige Jahre trennen uns von dem furchtbarsten Verbrechen, das die moderne Geschichte ~~Manner~~ aufzuweisen hat. Das Schicksal der überlebenden Opfer der deutschen Verfolgung legt Zeugnis davon ab, ~~dass~~ wie schwach das moralische Gewissen der Menschheit geworden ist.

Die heutige Versammlung ~~legt Zeugnis davon ab~~ zeigt, dass die besseren Menschen nicht Willens sind, das Furchtbare schweigend hinzunehmen. Diese Versammlung ist beseelt von dem Willen, dem menschlichen Individuum seine Würde und seine Rechte zu sichern. Sie steht für die Erkenntnis, dass eine erträgliches Dasein für die Menschen—ja eben das nackte Dasein—an das Festhalten an den ewigen moralischen Forderungen gebunden ist.

Für diese Haltung möchte ich ihr als Mensch und als Jude meiner Anerkennung und Dank aus.

Message on the dedication of Riverside Drive Memorial to the Victims of the Holocaust, New York, October 19, 1947

LEFT Einstein with Stephen Wise and Thomas Mann at the preview of the anti-war film *The Fight for Peace*, New York, October 5, 1938
© ACME PHOTO

EMBASSY OF ISRAEL
WASHINGTON, D.C.

שגרירות ישראל
ושינגטון

November 17, 1952

Dear Professor Einstein:

The bearer of this letter is
Mr. David Goitein of Jerusalem who is now serving as Minister
at our Embassy in Washington. He is bringing you the question
which Prime Minister Ben Gurion asked me to convey to you,
namely, whether you would accept the Presidency of Israel
if it were offered you by a vote of the Knesset. Acceptance
would entail moving to Israel and taking its citizenship.
The Prime Minister assures me that in such circumstances
complete facility and freedom to pursue your great scientific
work would be afforded by a government and people who are
fully conscious of the supreme significance of your labors.

Mr. Goitein will be able to give you
any information that you may desire on the implications of
the Prime Minister's question.

Whatever your inclination or decision
may be, I should be deeply grateful for an opportunity to
speak with you again within the next day or two at any place
convenient for you. I understand the anxieties and doubts
which you expressed to me this evening. On the other hand,
whatever your answer, I am anxious for you to feel that the
Prime Minister's question embodies the deepest respect which
the Jewish people can repose in any of its sons. To this
element of personal regard, we add the sentiment that Israel
is a small State in its physical dimensions, but can rise to
the level of greatness in the measure that it exemplifies
the most elevated spiritual and intellectual traditions which
the Jewish people has established through its best minds and
hearts both in antiquity and in modern times. Our first
President, as you know, taught us to see our destiny in these
great perspectives, as you yourself have often exhorted us to
do.

Therefore, whatever your response to
this question, I hope that you will think generously of those
who have asked it, and will commend the high purposes and
motives which prompted them to think of you at this solemn
hour in our people's history.

With cordial personal wishes,

Yours respectfully,

Abba Eban

Abba Eban

Professor Albert Einstein
Princeton, N.J.

The official offer of the presidency of the State of Israel. Letter
from Abba Eban, November 17, 1952.

I am deeply moved by the offer from our State of Israel, and at once saddened and ashamed that I cannot accept it. All my life I have dealt with objective matters, hence I lack both the natural aptitude and the experience to deal properly with people and to exercise official functions. For these reasons alone I should be unsuited to fulfill the duties of that high office, even if advancing age was not making increasing inroads on my strength. I am the more distressed over these circumstances because my relationship to the Jewish people has become my strongest human bond, ever since I became fully aware of our precarious situation among the nations of the world.

Einstein's response to the offer of the Israeli presidency, November 18, 1952

The Presidency of Israel

Upon the death of Chaim Weizmann, the first president of Israel, in November 1952, Einstein was proposed as his successor. The editor-in-chief of the Hebrew daily, *Ma'ariv*, Azriel Carlebach, initiated a public campaign urging that the Israeli government offer the presidency of the state to Albert Einstein. This proposal was taken up by Prime Minister David Ben-Gurion. The official approach was made by Abba Eban, Israel's Ambassador to the United States. At the time, Ben-Gurion is reported to have asked his aide, Yitzhak Navon, "Tell me what to do if he says yes! If he accepts, we are in for trouble." Hearing rumors of the offer, members of the Einstein household jokingly started to appoint ministers. Einstein was deeply moved by the offer but declined with sincere regret.

I also gave thought to the difficult situation that could arise if the government or the parliament made decisions which might create a conflict with my conscience; for the fact that one has no actual influence on the course of events does not relieve one of moral responsibility.

Einstein reveals further reasons for rejecting the post of Israeli president which he had not included in his official response. Letter to Azriel Carlebach, editor-in-chief, *Ma'ariv*, November 21, 1952.

The Hebrew University

Einstein envisioned the establishment of a Jewish spiritual center in Palestine. The core of that center would be a Jewish university in Jerusalem. Beginning in 1919, Einstein played a major role in discussions concerning the establishment of The Hebrew University. The main purpose of his first trip to the United States in 1921 was to raise funds for the university's planned medical faculty. During his only visit to Palestine in 1923, he gave the university's inaugural scientific lecture. That same year he edited the university's first scientific publication,

BELOW "The Mission of Our University." Message to The Hebrew University upon its official opening, April 1, 1925, published in *The New Palestine*, March 27, 1925.

which included an article by him and fellow scientist, Jakob Grommer.

From the twenties onwards, Einstein also played a central role in the development of the Jewish National & University Library: he allowed his name to be used for fund-raising and donated hundreds of books and periodicals to the library. Upon its opening in 1925, Einstein presented the university with the original manuscript of his general theory of relativity. He also served as a member of The Hebrew University's first Board of Governors and as chairman of its first Academic Council. Subsequently, major differences of opinion developed between Einstein and the chancellor, J. L. Magnes, concerning the running of the university. Einstein was opposed to the influence of American philanthropists on the handling of academic affairs—he advocated a German model whereby the academics would have the decisive say. In 1928, he resigned from his official posts at the university, yet remained closely involved in its affairs. In 1935,

The Mission of Our University

By ALBERT EINSTEIN

THE opening of our Hebrew University on Mount Scopus, at Jerusalem, is an event which should not only fill us with just pride, but should also inspire us to serious reflection.

A University is a place where the universality of the human spirit manifests itself. Science and investigation recognize as their aim the truth only. It is natural, therefore, that institutions which serve the interests of science should be a factor making for the union of nations and men. Unfortunately, the universities of Europe today are for the most part the nurseries of chauvinism and of a blind intolerance of all things foreign to the particular nation or race, of all things bearing the stamp of a different individuality. Under this regime the Jews are the principal sufferers, not only because they are thwarted in their desire for free participation and in their striving for education, but also because most Jews find themselves particularly cramped in this spirit of narrow nationalism. On this occasion of the birth of our University, I should like to express the hope that our University will always be free from this evil, that teachers and students will always preserve the consciousness that they serve their people best when they maintain its union with humanity and with the highest human values.

Jewish nationalism is today a necessity because only through a consolidation of our national life can we eliminate those conflicts from which the Jews suffer today. May the time soon come when this nationalism will have become so thoroughly a matter of course that it will no longer be necessary for us to give it special emphasis. Our affiliation with our past and with the present-day achievements of our people inspires us with assurance and pride *vis-à-vis* the entire world. But our educational institutions in particular must regard it as one of their noblest tasks to keep our people free from nationalistic obscurantism and aggressive intolerance.

Our University is still a modest undertaking. It is quite the correct policy to begin with a number of research institutes, and the University will develop naturally and organically. I am convinced that this development will make rapid progress and that in the course of time this institution will demonstrate with the greatest clearness the achievements of which the Jewish spirit is capable.

A special task devolves upon the University in the spiritual direction and education of the laboring sections of our people in the land. In Palestine it is not our aim to create another people of city dwellers leading the same life as in the European cities and possessing the European bourgeois standards and conceptions. We aim at creating a people of workers, at creating the Jewish village in the first place, and we desire that the treasures of culture should be accessible to our laboring class, especially since, as we know, Jews, in all circumstances, place education above all things. In this connection it devolves upon the University to create something unique in order to serve the specific needs of the forms of life developed by our people in Palestine.

All of us desire to cooperate in order that the University may accomplish its mission. May the realization of the significance of this cause penetrate among the large masses of Jewry. Then our University will develop speedily into a great spiritual center which will evoke the respect of cultured mankind the world over.

Einstein with Eliahu Elath, Israel's ambassador to the United States, May 1950. Elath served as president of The Hebrew University from 1962 to 1968.

PHOTO BY EDWARD H. ZWERIN; DONATED BY ELIAHU ELATH

FOREWORD

I am very happy to congratulate the Hebrew University on the attainment of its 25th anniversary. During this period much has been achieved, thanks to the devotion of the academic staff, in spite of difficulties both internal and external, of a lack of funds and two wars.

I can well envisage the University becoming increasingly important not only for the new State of Israel but also for the Jews throughout the world. But if it is indeed to become so then its spirit must keep pace with the greatness of the task. In other words, our highest ideal must be the acquisition and diffusion of knowledge. Only then can we create those permanent conditions in which practical achievements can also flourish and bring benefits to the country. A narrow, utilitarian spirit is as dangerous as is one which places undue emphasis on nationalism or on the purely formalistic observance of religious doctrines. We must also beware of the provincialism which so often accompanies baseless self-glorification.

I hope that the University will soon be able to resume its work undisturbed in its beautiful home, and that it will become a factor in Israel in strengthening the spirit of mutual understanding among men, which comes with selfless striving after truth.

ALBERT EINSTEIN

SCRIPTA
UNIVERSITATIS
ATQUE BIBLIOTHECAE
HIEROSOLYMITANARUM

MATHEMATICA ET PHYSICA
VOLUMEN I
CURAVIT A. EINSTEIN

HIEROSOLYMIS
MCMXXIII

Message on the occasion of the twenty-fifth anniversary of The Hebrew University's foundation, published in *The Hebrew University of Jerusalem 1925–1950* (Jerusalem, 1949)

LEFT Albert Einstein (ed.), *Scripta Universitatis atque Bibliothecae Hierosolymitanarum: Mathematica et Physica*, vol. 1 (Jerusalem, 1923). The first collection of scientific articles published by The Hebrew University.

Einstein's views were adopted by the university and he resumed his official ties.

During the several decades of his affiliation with the university, he stressed the institution's social and cultural importance for Israel and the Jewish people. As a lasting expression of this special association, he bequeathed his personal papers and literary estate to the university in his Last Will and Testament.

Einstein receives an honorary degree from The Hebrew University, at his office at the Institute for Advanced Study, Princeton, March 15, 1949 (with Leo Schwartz and High Salpeter of the American Friends of The Hebrew University)

PHOTO BY EDWARD H. ZWERIN

H. J. RES. 365

Mr. KEN...
the ...
print...

J...

Wherea...
sci...

Wherea...
in...

Wherea...
of...

Wherea...
lib...

1

2 *of*

3 That Albert Einstein is hereby unconditionally admitted to

4 the character and privileges of a citizen of the United States.

6 | Einstein and the United States

"Let me express my joy at being able to stay a while in this wonderful, sunny country among cheerful and friendly people and in constant contact with highly regarded scholars."

Albert and Elsa Einstein with U.S. President
Harding and members of the U.S. Academy
of Sciences, White House Lawn, April 1921

OPPOSITE, ABOVE Radio speech "A Greeting
to America" by Einstein, broadcast while on
board the SS *Belgenland,* December 11, 1930

On America

Like many European intellectuals, Einstein's attitude towards the United States was a mixture of
admiration, irreverence, and at times, bewilder-
ment. Following his first visit in 1921, Einstein
expressed his high regard for America's democratic
system, its scientific institutions and technology,
and its optimism, philanthropy, and organizational
capabilities. Yet he was critical of what he perceived
as the materialism and conformism of American
society, the commercialism of its culture, and its
political isolationism. In 1933, Einstein took up
residence in the United States as a refugee from
Germany. Congress proposed to grant Einstein
citizenship, but he rejected any preferential treatment. In 1935, Einstein decided to settle perma-

Radio-speech *Dec 1930*

S.S. BELGENLAND

A GREETING TO AMERICA.

This morning, when after an absence of ten years I once more am about to set foot upon the soil of the United States, the thought that is uppermost in my mind is this: this country, which through hard but peaceful labor has achieved a position of undisputed pre-eminence among the nations of the world, stands forth today as the one impregnable citadel of that ancient and high ideal of a political democracy.

Here in this land of yours, every man and woman is proud and jealous of his or her rights as citizens of a common country, and when I use the word "rights" I do not mean something vague and theoretical that stands writ upon a negligible piece of paper but something tangible and concrete that allows every member of the community, regardless of birth or background, to develop all his latent powers for the benefit of the community at large. Through the free development of brain and brawn your country has definitely proved that Liberty of the Individual is bound to be productive of much greater results than any sort of tyranny could ever hope to be, especially when all those who contribute to the ultimate results are not only filled with a justifiable pride in the State which they have helped to found and to maintain but also when they are willing and eager to bring certain occasional sacrifices for the benefit of the community at large.

And out of this consciousness of being in the truest sense of the word responsible for the welfare of the whole nation, it seems to me that that strange faith was born, that unswerving and unflinching faith of yours which believes in the great common interests of all peoples of the world and which has so often manifested itself in a concrete and practical form when it was necessary to further some cultural or scientific cause which was to be of ultimate benefit to the whole of society.

It is in your country, my friends, that those latent forces which eventually will kill the nefarious monster of professional militarism will be able to make themselves most clearly and definitely felt. Your political and economic position today is such that if ever you set your hand to this job in all seriousness, you can entirely destroy the dreadful tradition of military violence under which Europe, with its sad memories of the past, and to a certain extent the whole of the world, continues to suffer, even after the last dreadful warning of the Great War. It is along this line of endeavour that your mission lies at the present moment and should you be able and willing to accept this high duty which Fate has so unexpectedly placed in your hands, I know that

2

S.S. BELGENLAND

you will build yourself such a monument of gratitude and affection as never yet has been erected to any other country within the memory of man.

With this thought in my heart I greet you and I salute the soil which bears you in the happy expectation of renewing many old and cherished friendships and of widening my own point of view through the many interesting things I shall see during the time I shall be allowed to remain in your country.

(signed) Albert Einstein

nently in America and apply for citizenship. Five years later, on October 1, 1940, Einstein became an American citizen. During World War II, the FBI considered Einstein a risk to national security and objected to his receiving information on the progress of the A-bomb project. Nevertheless, the U.S. Navy asked Einstein to carry out some work for them on high explosives. Following the war, Einstein was critical of America's role in the nuclear arms race and warned of the dangers to democracy of McCarthyism and of unrestrained capitalism.

Einstein in his Princeton study with U.S. Navy personnel, carrying out research work for the U.S. Navy Bureau of Ordnance during World War II, July 24, 1943

COURTESY OF U.S. NAVY, NAVAL HISTORICAL FOUNDATION, HISTORICAL SERVICES

Einstein speaking at the dedication of the Pasadena Junior College astronomy building, February 26, 1931

PHOTO BY ROBERT SEARES / COURTESY CALTECH ARCHIVES

What first strikes the visitor with amazement is the superiority of this country in matters of technics and organization. Objects of everyday use are more solid than in Europe, houses infinitely more convenient in arrangement.... The second thing that strikes a visitor is the joyous, positive attitude to life. The smile of the people in photographs is symbolical of one of the American's greatest assets. He is friendly, confident, optimistic, and—without envy.

—Einstein on America, July 1921. Translation published in *The World As I See It* (New York, 1934).

THE WHITE HOUSE
WASHINGTON   25.I 34

In den Hauptstadt stolzer Pracht
Wo das Schicksal wird gemacht
Kämpfet froh ein stolzer Mann
Der die Lösung schaffen kann.

Beim Gespräche gestern Nacht
Herzlich Ihrer ward gedacht
Was berichtet werden muss –
Darum send' ich diesen Gruss

Albert Einstein.

In the Capital's proud glory
Where Destiny unfolds her story,
Fights a man with happy pride
Who solution can provide.

Of our talk of yester night
There are mem'ries bright;
In remembrance of our meeting,
Let me send you this rhymed
  greeting.

Poem by Einstein written for Queen Elisabeth of the Belgians during his stay at the White House on invitation of President Roosevelt, January 25, 1934. Translation by White House staff.

73D CONGRESS
2D SESSION

H. J. RES. 309

IN THE HOUSE OF REPRESENTATIVES

MARCH 28, 1934

Mr. KENNEY introduced the following joint resolution; which was referred to the Committee on Immigration and Naturalization and ordered to be printed

---

JOINT RESOLUTION

To admit Albert Einstein to citizenship.

Whereas Professor Albert Einstein has been accepted by the scientific world as a savant and a genius; and

Whereas his activities as a humanitarian have placed him high in the regard of countless of his fellowmen; and

Whereas he has publicly declared on many occasions to be a lover of the United States and an admirer of its Constitution; and

Whereas the United States is known in the world as a " haven of liberty and true civilization ": Therefore be it

1   *Resolved by the Senate and House of Representatives*
2   *of the United States of America in Congress assembled,*
3   That Albert Einstein is hereby unconditionally admitted to
4   the character and privileges of a citizen of the United States.

Joint Resolution of U.S. Congress granting Einstein U.S. citizenship, March 28, 1934

The New York Times.

NEW YORK, SUNDAY, APRIL 3, 1921. In Nine Parts, Including Rotogravure Picture Sections, Book and Magazine Section.

SYRACUSE TEACHER SLAYS COLLEGE DEAN AND KILLS HIMSELF

Instructor Had Brooded Over Impending Dismissal and Life of Continuous Failure.

PROF. WHARTON HIS VICTIM

Fires Five Shots Into Superior's Body in Latter's Office —Also Had Knife.

LEFT REMARKABLE LETTER

Wrote History of His Life in Which He Attributed Failures to Injustice.

Special to The New York Times.

SYRACUSE, N. Y., April 2.—Embittered by what he termed rank injustice from university authorities, Professor Holmes Beckwith, Instructor in Insurance, today shot and killed Dean J. Herman Wharton of the College of Business Administration, Syracuse University, and then ended his own life by the side of the man he had slain.

Beckwith fired five shots at Dean Wharton, all of them taking effect, and fired one bullet through his own brain from a large army revolver. Dean Wharton was shot in the right temple, twice in the region of the heart and twice in the legs.

The murder and suicide took place in the private office of Dean Wharton about 9 o'clock this morning, just after Beckwith had been admitted to a conference with Professor Wharton. The tragedy was not discovered until fifteen minutes later, presumably because carpenters were working on the outside of the office and stenographers and other office attendants believed the muffled sounds they had heard within the office were due to hammering.

In the last ten years Beckwith had been dismissed from nine positions, and he had been brooding over his affairs. He had been notified that his resignation from Syracuse University would be required, to take effect in June at the close of the college year, because he had failed to conduct his classes with the proper decorum. This he termed due to injustice and prejudice. He was especially bitter toward Dean Wharton. It has been learned, because he believed him to be responsible in part for his impending dismissal.

Thought Shots Were Hammering.

At 9:10 o'clock Professor Charles L.

Herrick Accepts Mission to France; Harvey Will Discontinue His Weekly

Special to The New York Times.

WASHINGTON, April 2.—Ex-Governor Myron T. Herrick of Ohio accepted today President Harding's tender of the office of Ambassador to France. At the President's invitation Mr. Herrick came to Washington last night and saw Mr. Harding today. No formal announcement of the tender and acceptance was made, but after Mr. Herrick's visit it became known that he would be nominated for the post when Congress assembles on April 11. In the meanwhile, the French Government will be asked if Mr. Herrick is acceptable—the usual procedure. He is popular in France and a cordial affirmative answer is expected.

There has been some doubt as to whether Mr. Herrick would accept the French mission. He has been credited with desiring to be appointed Ambassador to Great Britain, but President Harding had, from the beginning, made up his mind to appoint Colonel George Harvey to this office. For a time there were suggestions that General John J. Pershing would be asked to accept the Ambassadorship to France.

Because of his selection as Ambassador to the Court of St. James's Colonel Harvey is to discontinue the publication of Harvey's Weekly. With the Colonel overseas, he could not continue to give it the continuous personal touch that has been its chief asset, and many persons feel that it would be dangerous, if not unbecoming, for an Ambassador to be charged with responsibility for opinions that would appear in a publication of that kind. So Colonel Harvey intends, according to what he is telling friends in Washington, "to choke the baby to death."

This decision of Colonel Harvey is expected to cause a sigh of relief at the White House, the State Department and the London Foreign Office.

PROF. EINSTEIN HERE, EXPLAINS RELATIVITY

"Poet in Science" Says It Is a Theory of Space and Time, But It Baffles Reporters.

SEEKS AID FOR PALESTINE

Thousands Wait Four Hours to Welcome Theorist and His Party to America.

A man in a faded gray raincoat and a flopping black felt hat that nearly concealed the gray hair that straggled over his ears stood on the boat deck of the steamship Rotterdam yesterday, timidly facing a battery of cameramen. In one hand he clutched a shiny briar pipe and with the other clung to a precious violin. He looked like an artist—a musician. He was.

But underneath his shaggy locks was a scientific mind whose deductions have staggered the ablest intellects of Europe. One of his traveling companions described him as an "intuitive physicist" whose speculative imagination is so vast that it senses great natural laws long before the reasoning faculty grasps and defines them.

The man was Dr. Albert Einstein, propounder of the much-debated theory of relativity that has given the world a new conception of space, and time and the size of the universe.

Dr. Einstein comes to this country as one of a group of prominent Jews who are advocating the Zionist movement and hope to get financial aid and en-

CUBAN CONGRESSMAN SLAIN BY PARTY FOE

Quinones, National Leaguer, Is Shot Dead in Havana by Collado, Liberal.

CITY STIRRED BY TRAGEDY

Slayer Shot Antagonist Eight Times — Friends Had Only Just Dragged Them Apart.

Special Cable to The New York Times.

HAVANA, April 2.—On the famous Prado, in the heart of Havana, Fernando Quinones, a Conservative member of the Cuban Congress and recently National League candidate for Governor of Havana Province, was shot dead tonight by Ernesto Collado, a Liberal member of Congress from Las Villas and recently the party candidate for Governor of that province.

The shooting caused much excitement in the city and tonight groups of partisans are gathered discussing the affray.

A celebration of the National League victory in the election of Dr. Zayas as President was planned for this evening, but was suspended.

There is much talk as to the possible effect of the tragedy upon the session of Congress called for Monday to proclaim the election of Dr. Zayas.

Although he escaped immediately after the shooting Collado appeared later at Police Headquarters, two blocks from the scene of the killing, and surrendered.

BOTH SIDES FIRM IN BRITISH STRIKE; MANY MINES FILLING

Damage From Flooding in Staffordshire District Alone Is Put at £1,000,000.

HORNE ISSUES STATEMENT

Declares Miners' Action an Attempt to Intimidate the Government to Subsidize Industry.

STRIKERS' TEMPER UGLIER

Mine Officials in South Wales Are Forced to Withdraw Loyal Employes From Pits.

Special Cable to The New York Times.

LONDON, April 2.—Both sides stood firm today in the second day of the great coal strike, and neither made approaches to the other or any attempt to discover or explore avenues toward peace. A compromise at the moment, indeed, seems out of the question, although when Parliament meets on Monday there are sure to be many suggestions as to how the great calamity may be brought to an end.

Everyone is now watching to see what the railway men and transport workers will do when the Triple Alliance executives meet next week. The suggestion that has been made of a refusal on the part of the railway and transport men to handle coal, so as to nullify the advantage the coal owners possess in the existence of large stocks, is not accepted by those in the closest touch with the situation. The railwaymen, they say, will do nothing or everything. They will either leave the miners to fight their own battles or they will come out themselves. As to whether the latter is a real probability it is impossible to reach a conclusion.

The railway men are fairly satisfied, although uneasy over the numerous discharges that have occurred lately through depression of trade, and they have accepted the recent decline in wages, as a result of the fall in the cost of living, according to their sliding scale bargain. J. H. Thomas, too, is expected home tonight from the Continent, and his influence is counted on for peace.

Leaders More Radical Than Men.

But conclusions drawn from the attitude of the men and the entire absence of excitement in their general demeanor

First Visit

On his first trip to the United States in 1921, Albert Einstein accompanied Zionist leader Chaim Weizmann on a mission to raise funds for the planned medical faculty of The Hebrew University of Jerusalem. Even though Einstein had turned down previous invitations by scientific institutions to visit the United States, it was the cause of The Hebrew University which first brought Einstein to America's shores. Due to his newly acquired worldwide fame, Einstein was lionized throughout his trip—especially by the Jewish masses. On his arrival in New York, he was mobbed by a large group of reporters and photographers. Thousands cheered as his motorcade drove to City Hall, where he was welcomed by Mayor James J. Hylan. During

Menachem Ussishkin, Chaim and Vera Weizmann, Albert and Elsa Einstein, and Benzion Mossinson headed for the United States on board the SS *Rotterdam*, March 1921

© CENTRAL ZIONIST ARCHIVES, JERUSALEM

OPPOSITE Einstein's arrival in New York, as reported by the *New York Times*, April 3, 1921

© THE NEW YORK TIMES

Einstein in motorcade in New York,
April 4, 1921

OPPOSITE Letter from Einstein to his
friend Michele Besso, in which he expresses his opinions on his first impressions of the United States, May 28, 1921

his two-month tour of the United States, Einstein
held lectures at several significant scientific institu-
tions—Columbia University, the City College of
New York, Harvard University, the University of
Chicago, and the Case Institute of Technology in
Cleveland. Einstein also gave a series of four impor-
tant lectures on relativity at Princeton University. In
Washington, D.C., he addressed the National
Academy of Sciences and was received at the White
House by President Warren Harding. He also visit-
ed the Yerkes Observatory in Wisconsin.

America is interesting—with all its bustling activity, it is more capable of enthusiasm than other countries I usually "loiter" in. I had to let myself be shown around like a prize-winning ox, talk at countless large and small assemblies, hold innumerable scientific lectures. It's a miracle I could stand it all. But now it's over and what remains is the pleasant awareness that I've accomplished something really worthwhile and that I fought bravely for the Jewish cause in spite of all the protests from Jews and gentiles . . .

Einstein with large group at Yerkes
Observatory, Wisconsin, May 6, 1921

© YERKES OBSERVATORY, WISCONSIN, UNIVERSITY OF
CHICAGO

I am traveling with my wife to America on March 23 to
assist the Zionists with their fund-raising for the
university in Palestine. Therefore, I am forced to
postpone my lectures in Leiden and be absent from the
Solvay Conference. I regret this deeply, but one cannot
do everything, and I really think that my presence in
the United States will facilitate the "tapping" of the
dollar owners.

—From a letter from Einstein to physicist Paul
   Ehrenfest, in which he explains the reason for his trip
   to the United States, March 1, 1921

In California

When Einstein arrived in California for his first stay at Caltech in December 1930, he was immediately struck by the unfamiliar landscape en route to Pasadena, so different to the Central European countryside. The travel diaries he kept during his recurring sojourns in California in the early 1930s reveal that he fully enjoyed spending the winter months there. He described Pasadena as a "huge garden with grid-patterned streets" and was impressed by its surrounding hills and tropical

Albert and Elsa Einstein with Governor of California James Rolph, Jr., Pasadena, February 1931

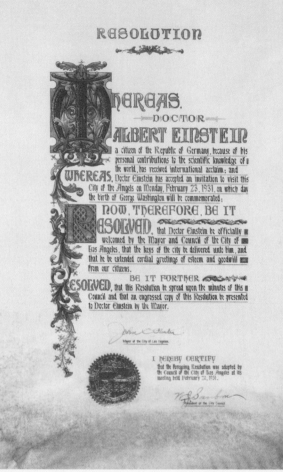

RESOLUTION

WHEREAS, DOCTOR ALBERT EINSTEIN a citizen of the Republic of Germany, because of his personal contributions to the scientific knowledge of the world, has received international acclaim; and

WHEREAS, Doctor Einstein has accepted an invitation to visit this City of the Angels on Monday, February 23, 1931, on which day the birth of George Washington will be commemorated;

NOW, THEREFORE, BE IT RESOLVED, that Doctor Einstein be officially welcomed by the Mayor and Council of the City of Los Angeles, that the keys of the city be delivered unto him, and that he be extended cordial greetings of esteem and goodwill from our citizens.

BE IT FURTHER RESOLVED, that this Resolution be spread upon the minutes of this Council and that an engrossed copy of this Resolution be presented to Doctor Einstein by the Mayor.

Mayor of the City of Los Angeles

I HEREBY CERTIFY that the foregoing Resolution was adopted by the Council of the City of Los Angeles at its meeting held February 20, 1931.

President of the City Council

Resolution officially welcoming
Einstein to the City of Los Angeles,
February 23, 1931

© CITY OF LOS ANGELES

vegetation. Einstein soon noticed the central role played by the automobile in the inhabitants' lives. He wrote that some cars may be "prehistoric," but that "everyone has one." He was also fascinated by self-service stores and the "ingenious" packaging of groceries. He thoroughly enjoyed traveling to Palm Springs, touring its surrounding desert landscape and interesting rock formations. Einstein also visited various film studios in Hollywood and attended special screenings of newly released films. Apart

from intense scientific discussions, he spent his time playing music, reading books on current issues, and lazing in the sun "like a crocodile." During his visits, Einstein also spoke at pacifist conferences and at fund-raising dinners on behalf of various causes.

Albert and Elsa Einstein in Palm Springs, January 1931

Here in Pasadena it's similar to paradise. Always sunshine and clean air, gardens with palm and pepper trees, and friendly people who smile at one and ask for autographs. Scientifically, it's very interesting and the colleagues are wonderful to me. But I am strongly drawn back to the rough North where one can walk barefoot and one is allowed to be coarse. In contrast, here everything is formal and respectable.

—From a letter by Einstein to the Lebach family on his stay in Pasadena, January 16, 1931

Speech by Einstein at Los Angeles dinner, February 1931

Engl. translation

The Ambassador
LOS ANGELES

Ladies and Gentlemen,

It is a distinguished pleasure for me, to be among you tonight, celebrating with a good glass of good, fresh water and to speak in this distinguished gathering of this country of sunshine. Yes, indeed, it is a true pleasure for me, to express my most greatful sentiments to you and to thank you for all that magnificent and elevating, which I witnessed in California.

I learned a good many many and exceptional things from clever and learned men, men, who deepen and further the sciences in progressiv quickness.

And yet I learned and achieved something still more difficult: Once I was a shy shy little man, who was unable, to bring out, to speak a single word publicly in a large gatherings. Well, I garthered, since I am here a dangerous knowledge of the art of oratory, yes and besides I learned to make my speech in a dignified pose manner and with a cordial smile at the same time I learned how to shake Hands with men and women.

For this reason don't be astonished, if I confess tonight, that I decided, to apply this art more profitable in the future. Yes, I decided for that reason for this reason I decided most firmly, at my next visit in the united states of America to be a candidate as president of the U.S.A.

But I am not in a great hurry. However, may I ask you now for your solemn promise, to vote for me.

For the time being, I thank you most cordially for the magnificient reception, you you gave me today tonight.

Wonderful sunrise from the lookout point above Untermayer's villa. Sun rises past the desert mountains and
bathes previously lead-gray rocks in golden sunlight
behind our lookout point. Then gloriously warm sun. Visit
to Gillette's cactus farm. Fantastic shapes there and—
photographers. In the afternoon, drive through the desert
to a ridge from where one descends to Palm Canyon, a
narrow valley floor covered with palm trees which is
bordered by bare rocks and scree.
—Excerpt from Einstein's U.S. travel diary on Palm Springs,
    January 26, 1931

Albert and Elsa Einstein with film director
Ernst Lubitsch (left), and their host Samuel
Untermayer, Palm Springs, January 1931
© BETTMANN/CORBIS

Einstein outside Fuld Hall, Institute for Advanced Study, Princeton, 1950s

PHOTO BY PERCY W. WITHERELL; COURTESY OF AIP EMILIO SEGRÈ VISUAL ARCHIVES

OPPOSITE Letter by Abraham Flexner to Einstein regarding the first meeting of the new Institute for Advanced Study, May 29, 1933

COURTESY OF THE ARCHIVES OF THE INSTITUTE FOR ADVANCED STUDY, PRINCETON

At Princeton

Einstein's ties with Princeton began in 1921 during his first trip to the United States, when he held a series of lectures at Princeton University. In the early 1930s, the noted educator Abraham Flexner secured funding for the establishment of a prestigious institute for research at Princeton. The aim of the planned Institute for Advanced Study was to enable eminent figures from the international scientific community to live and work in a peaceful and productive environment, free of any lecturing responsibilities. Following the Nazi rise to power in 1933, Einstein decided to emigrate to the United States and accept a permanent, lifelong position at the Institute in Princeton. Albert and Elsa Einstein initially resided at 2 Library Place. In 1935, they

~~100 EAST 42nd STREET~~
~~NEW YORK N. Y.~~

OFFICES
20 NASSAU STREET
PRINCETON
NEW JERSEY

May 29, 1933

Dear Professor Einstein:

Last week we moved our office from New York to 20 Nassau Street, Princeton. We have here extremely pleasant quarters, which will for the present answer very satisfactorily. Dean Eisenhart has arranged quarters for you and the other members of the Institute in Fine Hall. I am sure that you will be very happy there within easy reach of anything that you need in the way of books or other facilities. We have thus far accepted eight or ten advanced workers, all of whom are capable of pursuing their own work with such informal direction as members of the Institute may find it worth while to give. Almost without exception these workers have already held university posts in this country or abroad and are coming to the Institute for a year in order to carry forward their own work. You will see, therefore, that we are doing what we started out to do, namely, getting together a group of distinguished persons, placing them under ideal conditions for their own work and associating with them in an informal way younger men who may from time to time enjoy advice and aid.

In order to have some common understanding, I should like to have a meeting of the members of the Institute on the morning of Monday, October 2. I have myself no plans to suggest, but I should like to bring the members together in order to listen to them while they consider the different ways in which they propose to carry on their work.

I imagine that you are now at Christ Church, and I envy you your sojourn

---

Professor Einstein   May 29, 1933   2

there, but Princeton is really very beautiful at this time of year, quiet, cool, and full of birds and flowers and trees in full blossom.

I am writing Mrs. Einstein by this same mail about some details regarding which I need not bother you.

With all good wishes and very high regard,

Sincerely yours,

Abraham Flexner

P.S.
I saw Ladenburg yesterday afternoon who read me many letters from German colleagues, not one of them a Jew, but all deploring bitterly the terrible things that are happening in Germany.
A.F.

Herrn Prof. Dr. Albert Einstein
Villa Savoyarde
Coq sur Mer, near Ostend, Belgium

AF:ESB

---

purchased a house close by at 112 Mercer Street. At first, the Institute was located at Princeton University's Fine Hall; from 1939 onwards, Einstein had an office at Fuld Hall on the Institute's own campus. Einstein was obviously attracted to Princeton by its academic and intellectual life and by its quiet beauty. Einstein officially retired in 1945, yet continued his research at the Institute until his death in 1955. Throughout his years in Princeton, Albert Einstein was the town's most celebrated inhabitant.

> In view of all the difficult things I have experienced these last few years, I am all the more grateful that I have been given a workplace and a scientific environment at Princeton University, which could not conceivably be better or more harmonious.
> —Letter by Einstein to President H. W. Dodds, Princeton University, following the death of Elsa Einstein, January 14, 1937

Princeton. 20. XI. 33

[handwritten letter in German]

Liebe Königin!

[handwritten text]

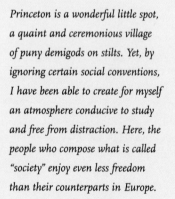

Princeton is a wonderful little spot, a quaint and ceremonious village of puny demigods on stilts. Yet, by ignoring certain social conventions, I have been able to create for myself an atmosphere conducive to study and free from distraction. Here, the people who compose what is called "society" enjoy even less freedom than their counterparts in Europe.

In Princeton, we have settled down really well. The place is charming, altogether different from the rest of America. The house in which we live is especially beautiful—very large, airy, comfortable, exceptionally well situated right in the center of old park grounds. Here, everything is tinged with Englishness— downright Oxford style.

—Letter by Elsa Einstein to Hertha Einstein on settling in at 2 Library Place, Princeton, February 24, 1934

Einstein at Princeton, 1950s

OPPOSITE, ABOVE Letter by Einstein to Queen Elisabeth of the Belgians on Princeton, November 20, 1933. Translation published in *Einstein on Peace* (New York, 1960).

OPPOSITE, BELOW Einstein's home at 112 Mercer St., Princeton

April 22, 1947

Dear Friend:

      I write to you for help at the suggestion of a friend.

      Through the release of atomic energy, our generation has brought into the world the most revolutionary force since prehistoric man's discovery of fire. This basic power of the universe cannot be fitted into the outmoded concept of narrow nationalisms. For there is no secret and there is no defense; there is no possibility of control except through the aroused understanding and insistence of the peoples of the world.

      We scientists recognize our inescapable responsibility to carry to our fellow citizens an understanding of the simple facts of atomic energy and their implications for society. In this lie our only security and our only hope - we believe that an informed citizenry will act for life and not for death.

      We need $1,000,000 for this great educational task. Sustained by faith in man's ability to control his destiny through the exercise of reason, we have pledged all our strength and our knowledge to this work. I do not hesitate to call upon you to help.

                  Faithfully yours,

                  A. Einstein.

American Scientists

Various American scientists played a significant role in the verification and development of Einstein's scientific theories. The Michelson-Morley experiment of 1887, which attempted to measure the relative motion of the Earth through a hypothetical ether, preceded Einstein's conclusion in 1905 that the speed of light is a universal constant. In 1916, Robert A. Millikan provided an experimental verification of Einstein's equation of the photoelectric effect. Millikan was also instrumental in inviting Einstein to California in the early 1930s, where he held important discussions on relativity with Caltech's cosmologists. American astronomers also influenced Einstein's theories: William W. Camp-

bell provided one of the crucial verifications of general relativity, and Edwin P. Hubble's work led to Einstein's revision of his theory of a cosmological constant. In the 1930s, Einstein played a major role in finding employment for refugee European scientists in the United States. This influx of talented émigrés had an important impact on the development of science in the United States during and after the Second World War. During his years at Princeton, he continued his quest for a unified field theory with various assistants at the Institute for Advanced Study.

Einstein and other "modern pioneers in science" who received citations at the Copernican Quadricentennial, Carnegie Hall, New York, May 24, 1943. The recipients included John Dewey, Walt Disney, Henry Ford, and Orville Wright.

OPPOSITE, ABOVE Letter by Einstein on behalf of the Emergency Committee of Atomic Scientists, Inc., April 22, 1947

OPPOSITE, BELOW Albert Einstein and J. Robert Oppenheimer, Princeton, late 1940s
ALFRED EISENSTAEDT / TIMEPIX

I have a warm admiration for the achievements of
American institutes of scientific research. We
[Europeans] are unjust in attempting to ascribe
the increasing superiority of American research-work
exclusively to superior wealth; zeal, patience, a spirit
of comradeship, and a talent for co-operation play
an important part in its successes.
—Einstein on American scientists, July 1921.
  Translation published in *The World As I See It*
  (New York, 1934).

ABOVE LEFT Honorary diploma of the National Academy of Sciences of the United States of America, which states that Einstein has been elected a foreign associate of the Academy, April 26, 1922

In Edison, one of the great technical inventors to whom we owe the possibility of alleviation and embellishment of our outward life has departed from us. An inventive spirit has filled his own life and our existence with bright light. Thankfully we accept his legacy, not only as a gift of his genius, but also as a mission placed in our hands.

—Obituary by Einstein for Thomas Alva Edison, October 19, 1931. Translation published in the *New York Times*.

LEFT Letter by Robert A. Millikan to Einstein, informing him of his election to the National Academy of Sciences, May 22, 1922

COURTESY OF THE ARCHIVES, CALIFORNIA INSTITUTE OF TECHNOLOGY

OPPOSITE Einstein at Mount Wilson Observatory with Director Walter S. Adams and astronomer William Wallace Campbell, January 1931

© HULTON-DEUTSCH COLLECTION / CORBIS

Einstein with a group of colleagues at Mount Wilson Observatory, including Walther Mayer, Charles E. St. John, W. S. Adams, and Edwin P. Hubble, 1932

COURTESY OF THE ARCHIVES, CALIFORNIA INSTITUTE OF TECHNOLOGY

OPPOSITE Einstein with fellow physicists Albert A. Michelson, Robert A. Millikan, W. S. Adams, Walther Mayer, and M. Ferrand in front of the Athenaeum at Caltech, January 7, 1931

At Caltech

Einstein spent the winter semesters of three consecutive years—from 1931 to 1933—at the California Institute of Technology in Pasadena. He was invited to serve as visiting professor at Caltech by Robert A. Millikan, the institute's first president. Einstein was attracted to Caltech for a number of reasons. Millikan had assembled an impressive group of theoretical physicists such as Richard C. Tolman and Paul S. Epstein at the Institute, with whom Einstein could discuss various topical issues of relativity as well as his ongoing quest for a unified field theory. The close proximity to Caltech of Mt. Wilson Observatory enabled Einstein to confer with astronomers Charles E.

Expect Great Results From Meetings—Meantime Interest Centres in Big Telescope.

Special Correspondence, THE NEW YORK TIMES.

LOS ANGELES, Dec. 11—Southern California is becoming the universal peep hole for the gimlet eye of science. Strange things are going forward at the Institute of Technology where Dr. Millikan makes the atoms and the molecules lie down, roll over and say "Uncle." The visit of Professor Einstein is being anticipated by the savants at the Pasadena Research Institution with an eagerness akin to that of a small boy faced by a mysterious looking package on the day before Christmas. Something portentous may come out of the impending huddles into which the world's mightiest men of science are preparing to engage.

Just now public interest is centring in the locating of the world's greatest telescope, the 200-inch reflecting instrument, which is expected, figuratively, to bring the earth within speaking distance of the moon and some of the other planets. One of the locations is on Table Mountain 7,500 feet high, easily accessible from the university where the laboratory work will be done and 1,500 feet higher than Mount Wilson where the present 100-inch instrument, now the world's largest telescope, is located. The other site is on Palomar Mountain in San Diego County, with an altitude of 6,100 feet, comparable with Mount Wilson, seventy-five miles from the institute in an isolated section, and eight miles from the nearest village.

If the latter site is selected a large amount of road building will be required before the job of housing and transporting the delicate mechanism and priceless lens can be undertaken.

Einstein's expected arrival in California, as reported by the *New York Times*, December 11, 1930

© THE NEW YORK TIMES

St. John and Edwin P. Hubble, who had just discovered the red shift effect and concluded that the universe was expanding. In addition, the political situation in Germany in the early 1930s was becoming increasingly precarious for prominent Jews like Einstein. Millikan disapproved of the political statements on pacifism, Nazism, and American domestic affairs made by Einstein during his visits to the Institute. Nevertheless, Caltech's board invited Einstein to accept a permanent position at the Institute, but Einstein declined.

GEORGE ELLERY HALE
PASADENA, CALIFORNIA

Mount Wilson Observatory
March 6, 1933

Professor Dr. Albert Einstein
The Athenaeum
Pasadena.

Dear Professor Einstein:

Our two new determinations of the general magnetic field of the sun have advanced more slowly than I had hoped, but Dr. Langer already seems fully convinced of the validity of our earlier results and Dr. Strong's very interesting measures by an entirely different method will probably yield fairly definite conclusions by tomorrow. Meanwhile, the spectrograph has been prepared to obtain additional photographs with new quarter-wave plates, having mica strips twice as wide as the old ones. With these plates we hope to get more perfect intensity-curves and therefore better measures of displacements with the Zeiss microphotometer.

I have given further thought to the international problem, and would like to discuss it with you if you are out for a walk and can conveniently call at my laboratory. I am here daily from 10 to 12.30 in the morning and 2.45 to 5 in the afternoon, though on Thursday I must go to see my physician in Los Angeles at 2 and cannot be back here until 4 o'clock. If you are too crowded for time I will write again, but a short talk would be much more useful.

With warmest regards to you and Mrs. Einstein, in which my wife joins,

Yours very cordially,

George E. Hale

GEH:G

No doubt this is the most wonderful offer which has ever been made to a scholar; and the prospect of life in your congenial circle among splendid colleagues and in this sunny and healthy country is extremely enticing. Nevertheless, I cannot accept the offer in its current form for the following reason: as a fifty-year-old, one should no longer entirely change one's human surroundings. It would mean an upheaval in one's life, which could not be compensated for by anything else.

Letter by astronomer George Ellery Hale to Einstein regarding the general magnetic field of the sun, March 6, 1933
COURTESY OF THE ARCHIVES, CALIFORNIA INSTITUTE OF TECHNOLOGY

LEFT Letter by Einstein to Arthur H. Fleming, Trustee of Caltech, in which Einstein declines offer of permanent position at Caltech, 1931

OPPOSITE Astronomer Charles E. St. John shows Einstein the red shift in the sun's spectrum at the Mt. Wilson Observatory near Pasadena, 1931
PHOTO BY FERDINAND ELLERMAN; COURTESY OF AIP EMILIO SEGRÈ VISUAL ARCHIVES

My dear Colleague

I am deeply grateful to you for your very generous letters. They only made me the more sad that circumstances were so cruel against us, in preventing us from having a real talk, aus den Herzen, during your visit. What is past is past. But I am abidingly grateful for the complete candor and heartiness of your letters. It would have been too cruel were it left otherwise between us —

the flickers of the unpleasant past.

You will again come to the United States, or we to Germany, & then we shall talk.

Please give my respects to Frau Einstein.

With high regard

Very sincerely

Felix Frankfurter

Letter by Harvard Law Professor (and future Supreme Court Justice) Felix Frankfurter to Einstein on his visit to the United States, July 1, 1921 (pp. 1 and 5)

It was in America that I first discovered the Jewish people. I have seen any number of Jews, but the Jewish people I had never met either in Berlin or elsewhere in Germany. This Jewish people, which I found in America, came from Russia, Poland, and Eastern Europe generally. These men and women still retain a healthy national feeling; it has not yet been destroyed by the process of atomization and dispersion.

—Einstein on his first encounter with the Jewish masses in the United States, July 1921. Translation published in *About Zionism* (London, 1930).

American Jewry

Einstein's first encounter with American Jewry took place during his visit to the United States in 1921. This fund-raising trip was the earliest instance in which his fame was utilized to raise money and exert influence in the United States on behalf of Jewish issues. Einstein was very scrupulous about the Jewish causes to which he was willing to lend his name. He was particularly interested in the consolidation of the Jewish community in Palestine, the education of Jewish youth, and the strengthening of Jewish social solidarity. In the 1930s, Einstein occasionally participated in public

Einstein with Meyer Weisgal, secretary general of the U.S. section of the Jewish Agency, and Einstein's secretary, Helen Dukas, at the Anglo-American Committee of Inquiry on Palestine, Washington, January 11, 1946

RIGHT Einstein with Rabbi Stephen Wise and New York Mayor Fiorello La Guardia at dinner in honor of Wise's sixtieth birthday, New York, March 1934

© BETTMANN/CORBIS

BELOW Letter by U.S. Supreme Court Justice Louis D. Brandeis to Einstein regarding the establishment of an American Jewish university, March 21, 1935

COURTESY OF FRANK B. GILBERT AND ALICE B. POPKIN

OPPOSITE, TOP LEFT "Fan letter" from Rabbi Stephen Wise to Einstein, December 12, 1945

COURTESY OF AMERICAN JEWISH HISTORICAL SOCIETY AND THE AMERICAN JEWISH CONGRESS

OPPOSITE, RIGHT Speech by Einstein at the Los Angeles Jewish Community Testimonial Dinner, winter 1931

rallies and gave radio addresses on behalf of various Jewish and Zionist organizations. With the rise of Nazism, he labored tirelessly to aid Jewish and political refugees, facilitating their immigration to the United States and helping them to obtain suitable employment. Einstein was strongly opposed to assimilationist groups among U.S. Jewry and warned them not to follow what he saw as the negative example of German Jewry. Einstein himself became a frequent target for American anti-Semites: as a prominent immigrant and outspoken critic of McCarthyism and the cold war, he received a fair share of hate mail. In his later years, he refrained from public appearances and preferred to issue statements and publish short articles on topical Jewish subjects.

Dr. Wise's personal
address is 40 W 68 St.
December 12, 1945.

Professor Albert Einstein
112 Mercer St
Princeton, N.J.

Dear friend:

I do not know whether you are enough of an American
to understand what the term "fan letter" means. It
means a letter of appreciation and gratitude written
to one for some great service. This is frankly and
unashamedly my fan letter to you, dear Professor Einstein.

You stated our Zionist case in your address before
the physicists with such clarity and wisdom that,
instead of speaking last night before a great Zionist
gathering, I rose and read your statement published in
PM. A thousand thanks to you! Your evaluation of the
situation as it obtains today must be of enormous help
to us. With all my heart's gratitude and affection,
I thank you and thank you!

Ever yours,

SSW:S

Among all those whom I have personally met who have labored in the cause of justice and in the interest of the hard-pressed Jewish people, only a few were at all times selfless—but there was no one who gave his love and energy with such consuming devotion as Stephen Wise.... By relentlessly exposing the weakness and imperfections both in our own ranks and in the larger political arena of the non-Jewish world, he has made great and lasting contributions wherever he has gone.
—Greeting by Einstein to Rabbi Stephen Wise on his seventy-fifth birthday. Translation published in *Opinion* (March 1949).

Let me express my joy at being able to stay a while in this wonderful, sunny country among cheerful and friendly people and in constant contact with highly regarded scholars. I am also delighted with the cordial reception recently bestowed upon me by the Jewish community. Please accept my sincere thanks.

During the 1940s

7 | Einstein at Leisure

"I like sailing because it is the sport which demands the least energy."

Einstein playing his violin on board the *Belgenland*, January 1931

Music

In music, Einstein found relaxation from the strains of his theoretical work and his public activities. Music provided him with "the highest possible degree of happiness." Sometimes, the solution to a problem in physics would come to him while playing music.

The characteristics he appreciated in music were purity, clarity, simplicity, and balance.

Mozart was his ideal composer. He also enjoyed Baroque composers such as Bach, Vivaldi, and Corelli. He liked Schubert and early Beethoven but disliked Wagner.

Reclams Universum

ILLUSTRIERTE WOCHENSCHRIFT

FERNRUF 24930
TELEGRAMME: RECLAM LEIPZIG

SCHRIFTLEITUNG   LEIPZIG, 24. März 1928
                                         INSELSTRASSE 22/24

R.

Herrn

Professor Dr. Albert Einstein,

Berlin W 30,

Haberlandstr. 5

Sehr verehrter Herr Professor!

Dürfen wir Sie an unser Schreiben vom 20. v. M. erinnern, auf das wir leider noch ohne Nachricht geblieben sind? Aus redaktionellen Gründen wäre uns eine baldige Benachrichtigung sehr erwünscht, und wir hoffen gern, daß Sie den erbetenen Artikel für uns schreiben werden.

Wir verbleiben mit dem Ausdruck unserer ausgezeichneten Hochachtung

ergebenst

Schriftleitung von
Reclams Universum.

Was ich zu Bach's Lebenswerk zu sagen habe; Hören, Spielen, verehren und — das Maul halten!

Reply to a query of the illustrated weekly *Reclams Universum* on the music of Johann Sebastian Bach, March 24, 1928

Einstein with group of musicians in Princeton, New Jersey, November 1933

© NEW YORK TIMES PICTURES

Wolfgang Amadeus Mozart, *Three Divertimenti Arranged for Piano and Violin.* From Einstein's extensive collection of musical scores.

In popular mythology, Einstein is famous for his violin. He took lessons between the ages of six and fourteen but resented his teachers' system of learning by rote. Only at the age of thirteen did he discover the beauty of music through Mozart's sonatas. In his later years, he claimed that if he had not been a scientist he would have become a musician. He often played chamber music with friends and associates. Opinions vary as to the quality of his playing. He was probably a good amateur who had an intuitive understanding of music.

Towards the end of his life, he grew to dislike the "scratching" sounds he produced on the violin and he eventually abandoned the violin for improvisations on the piano.

Johann Sebastian Bach, *Sonatas for Violin and Harpsichord* (vol. 1), Gamut Records. From Einstein's extensive record collection.

OPPOSITE, ABOVE Herbert Katzman (left), a director of the America-Israel Cultural Foundation, and Menahem Avidom, the Israeli composer, visiting Einstein at his Princeton home, January 15, 1951. Katzman observed that Einstein's violin was not very well kept. "I know, I know," Einstein replied.

PHOTO BY PHILIPPE HALSMAN; © HALSMAN ESTATE

DRAWING BY DAVID LEVINE; © 1983 NYREV, INC.

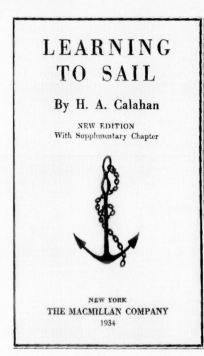

LEARNING
TO SAIL

By H. A. Calahan

NEW EDITION
With Supplementary Chapter

NEW YORK
THE MACMILLAN COMPANY
1934

H. A. Calahan, *Learning to Sail* (New York, 1934). A book from Einstein's personal library.

At Watch Hill, Rhode Island, 1934

Sailing

Einstein's favorite sport was sailing. In Berlin, he sailed on the lakes around the city. Einstein kept his boat *Tümmler* (German for "porpoise") at his summer house at Caputh outside Berlin. In the United States, he enjoyed sailing his boat *Tinef* (colloq. German for "worthless thing") throughout his summer holidays in New Jersey, New York, and Rhode Island.

Sailing allowed Einstein to lose himself in thought while the wind carried him along. He was not interested in speed or competition. He was delighted when there was a lull and the boat came to a standstill or ran aground. He would often keep a notebook at hand, scribbling away at scientific calculations when the sea was calm. Intriguingly, Einstein could not swim, yet stubbornly refused to carry a life jacket or an emergency motor on board.

Wir zwei und Tinef grüssen Sie
Bevor wir und ersoffen
Bis es soweit, verlass' uns nie
Freundschaft und frohes Hoffen.

Frau Damman zu Weihnachten 1944
A, Einstein.

Both we and Tinef send regards
As you see, we have not drowned
Until such time is on the cards
May friendship and hope abound.

On an accident with his boat *Tinef* in the summer of 1944. Quatrain dedicated to Ruth Damman, Christmas 1944. Translation: Ze'ev Rosenkranz and Susan Worthington.

At Huntington, Long Island, 1937

© LOTTE JACOBI COLLECTION, UNIVERSITY OF NEW HAMPSHIRE

8 | Correspondence with Children

"Dear Mr. Einstein, I saw your picture in the paper. I think you ought to have your haircut, so you can look better."

Correspondence with Children

Einstein was a subject of fascination for children all around the world. They sent him letters asking all kinds of questions and some even offered very practical advice. Many enclosed drawings, photos, and small gifts. Einstein obviously enjoyed receiving these letters, as he kept a large number of them. His carefully worded responses reveal his wish to foster children's natural inquisitiveness, his great affection for them, and his quirky sense of humor.

In his old age, walking around Princeton, he would always greet children and babble with the babies. Children were baffled and intrigued by the eccentricity of Einstein's appearance: the unruliness of his white hair, his socklessness, his baggy coat and woollen cap.

Postcards sent from Jewish schoolchildren in Berlin on the occasion of Einstein's fiftieth birthday, March 14, 1929

LEFT Einstein in his Princeton home with Jewish refugee children who survived the Holocaust, March 1949

PHOTO BY PHILIPPE HALSMAN; © HALSMAN ESTATE

BELOW Einstein's astronomic instruments presented by him to the Ben Shemen children's village, Israel, December 1928

OPPOSITE Drawings from Japanese schoolchildren for Einstein's seventieth birthday, Takatsumi Primary School, Osaka Prefecture, March 1949

Baby John Steidig sitting on Einstein's lap, Deep Creek Lake, Maryland, September 1946

John Jurgensen suggests a flight to Mars or Venus, 1952

OPPOSITE Einstein holding five-month-old James William Pietsch outside his Princeton home, July 13, 1951

Dear Dr. Einstein

My Father and I are going to build a rocket and go to Mars or Venus. We hope you will go too! We want you to go because we need a good scientist and some one who can guide a rocket.

Do you care if Mary goes too? She is two years old. She is a very nice girl.

Everybody has to pay for his food because we will go broke if we pay!

I hope that you have a nice trip if you go.

Love
John Jurgensen
Culver,
Indiana

June 19, 1951

Miss Monique Epstein
381 Central Park West
New York City

Dear Monique:

There has been an earth since a little
more than a billion years. As for the question of the
end of it I advise: Wait and see!

Kind regards,

A. Einstein.

Albert Einstein

I enclose a few stamps for
your collection.

Reply to an inquisitive child's query
regarding the longevity of our planet,
June 19, 1951

March 26, 1955

To the 5th Grade
Farmingdale Elementary School
Farmingdale N.Y.

Dear Children,

I thank you all for the birthday gift
you kindly sent me and for your letter of congratulation
Your gift will be an appropriate suggestion to be a litt
more elegant in the future than hitherto. Because neckti
and cuffs exist for me only as remote memories.

With kind wishes and regards,

yours sincerely,

A. Einstein.

Albert Einstein.

Einstein thanks the 5th grade class of
Farmingdale Elementary School, New York,
for their birthday present, March 26, 1955

1951

DEAR MR. EINSTEIN
I AM A LITTLE GIRL OF
SIX.
I SAW YOUR PICTURE
IN THE PAPER.
I THINK YOU OUGHT TO
HAVE YOUR HAIRCUT,
SO YOU CAN LOOK
BETTER.
CORDIALLY YOURS,
ANN G. KOCIN.

A "practical" suggestion from six-year-old
Ann G. Kocin, 1951

H

Pro
Ins
Pri

Dea

In
in
leather tanners, manufacturers of mens and wor
and retail merchants and buyers, we are asking
favor, to start a collection of shoes worn by
persons.

The Shoe Club feels that a collection of shoes
been worn by men of renown will be an inspirat
younger members of the shoe industry not only
manship, but to show them that their livelihoo
service to mankind that they can be proud of.

Would you be kind enough therefore, to send us

9 | The Curiosity File

"Dear Professor,
Would it be reasonable to
assume that it is while a person
is standing on his head—or
rather upside down—he falls
in love and does other foolish
things?"

69–73 Fifty-eighth avenue,
Maspeth, L.I.,
New York City, N.Y.,USA

198

Professor Einstein,
c/o Commander Oliver Locker-Lampson,
4 North Street, Westminster,
London, S.W.1, England.

Dear Professor,

    I am sorry I cannot express this well enough in German.

    I understand the world moves so fast it, in effect, stands still, or so it appears to us. Part of the time it seems a person is standing right side up, part of the time on the lower side he is standing on his head, upheld by the force of gravity, and part of the time he is sticking out on the earth at *right* angles and part of the time at left angles.

    Would it be reasonable to assume that it is while a person is standing on his head — or rather upside down — he falls in love and does other foolish things?

                    Yours truly,

                    Frank Wall

Falling in love is not at all the most stupid thing that people do—but gravitation cannot be held responsible for it.

A rather idiosyncratic view of Einstein's theory of gravitation. Letter from Frank Wall, New York, 1933, and Einstein's drafted response.

The Curiosity File

Einstein's mythic stature brought with it a huge and varied correspondence. Einstein and his secretary, Helen Dukas, relished curious and eccentric items and preserved them in the "komische Mappe" (the "curiosity file"). Some seven hundred items have been preserved with a special section for curious envelopes. Only very rarely does one find a reply to these unusual letters in the Einstein Archives. The files include fan mail, marriage proposals, eccentric suggestions, and weird scientific theories. A less welcome side to these files is the hate mail sent to Einstein, usually anti-Semitic and anti-leftist in nature.

The SHOE CLUB, Inc.

HOTEL McALPIN
NEW YORK

TELEPHONE:
PENN 6-5700

February 7th
1936

Professor Albert Einstien
Institue for Advanced Studies
Princeton, N. J.

Dear Professor Einstien:

In behalf of the Shoe Club, the outstanding organization
in the shoe industry comprising a membership of 400
leather tanners, manufacturers of mens and womens shoes
and retail merchants and buyers, we are asking an unusual
favor, to start a collection of shoes worn by outstanding
persons.

The Shoe Club feels that a collection of shoes that have
been worn by men of renown will be an inspiration to the
younger members of the shoe industry not only in crafts-
manship, but to show them that their livelihood is of a
service to mankind that they can be proud of.

Would you be kind enough therefore, to send us one of
your old shoes, preferably the right shoe, and also fill
out the enclosed autograph? This shoe will be permanently
preserved and on display in our headquarters at the Hotel
McAlpin, New York City.

Thanking you in advance for complying with this request,
I am,

Very truly yours,
THE SHOE CLUB INC.

F. J. Murray
Executive Secretary

FJM:SH

A request for Einstein's right shoe. Letter
from F. J. Murray, The Shoe Club, Inc., New
York, February 7, 1936.

At Saranac Lake, New York, 1936

Mr. Einstein
The atomic scientist
U. S. A
R 435
GRAM
1 rupiah
INDONESIA
AIR
INDONESIA 7 1/2 SEN
DEV. No. 1/BT.

DE LAND DEC 1 4 PM 1953 FLA.
UNITED STATES POSTAGE
3 CENTS 3
EINSTEIN
U. S. A.

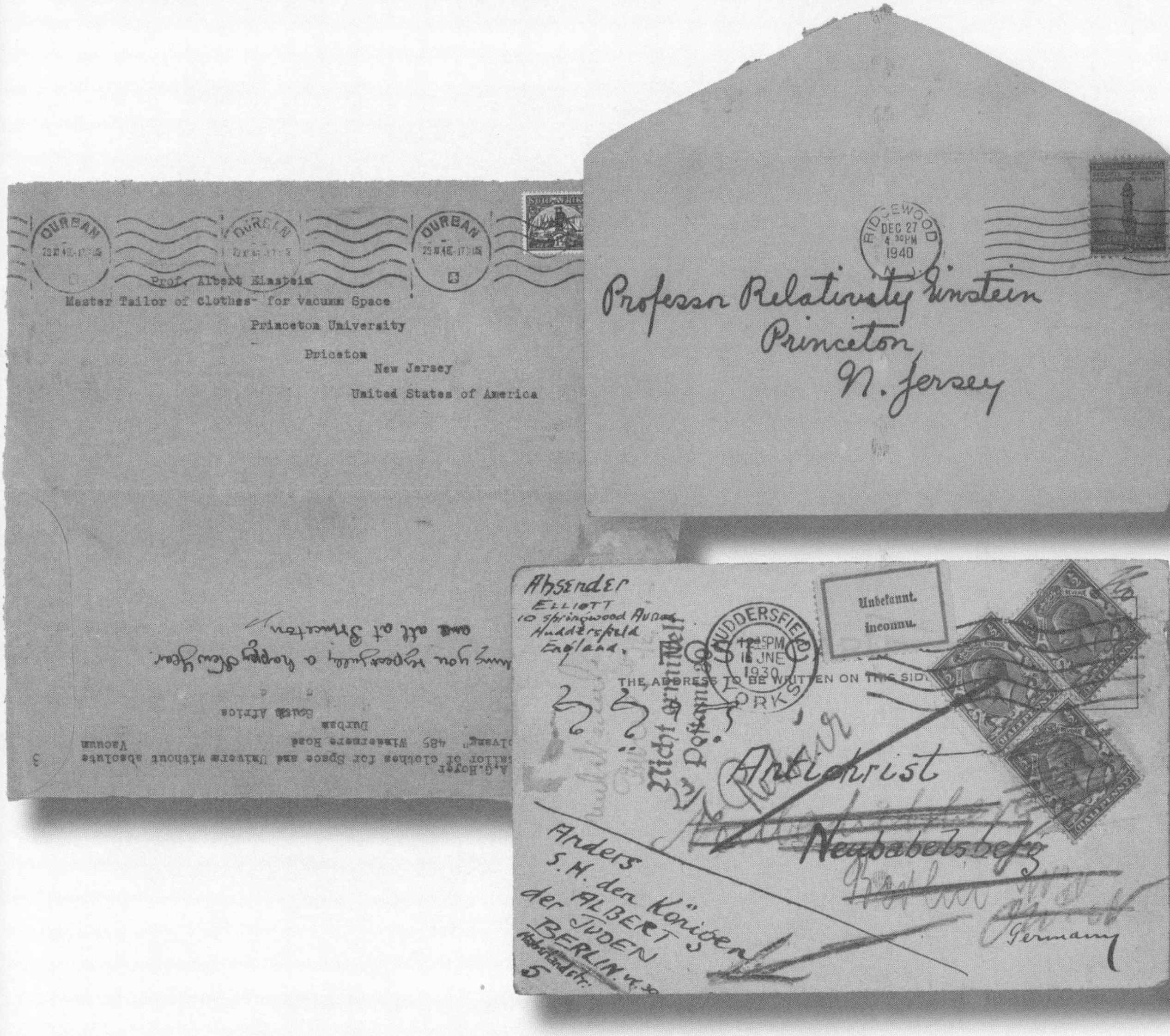

Envelopes from the "curiosity file"

Albert and Elsa Einstein with a group of Hopi at the Grand Canyon Reservation, February 28, 1931

PHOTO BY FRED HARVEY; © NEW YORK TIMES PICTURES

LEFT The quirky professor arriving in London at Victoria Station, 1933

PHOTO BY LONDON EXPRESS

OPPOSITE At the home of Ben Meyer in Santa Barbara, California, February 18, 1933

COURTESY OF THE ARCHIVES, CALIFORNIA INSTITUTE OF TECHNOLOGY

S.S. John M.Schofield, at sea,
Proceeding from Brake,Aldenberg,Germany,
to New York City, August 3, 1946.

Professor Albert Einstein,
Princeton University,
Princeton, New Jersey.

My dear Professor Einstein:

Forgive me if I seem presumptuous in addressing you, but I
have heard of your keen sense of humor and believe that the following true
anecdote may give you a friendly smile.

Our ship has just delivered a cargo of some 309,000 bushels
of relief wheat to the grain elevators at the village of Brake, Port of Bremen.
Our last night the Boatswain and the Carpenter, two old time sailors came back
from shore a little the worse for wear, a little kitten about six weeks old.
When they sobered up the next morning they decided to become joint foster parents
of the kitten. The poor kitten must have been getting a meagre diet and seemed weak
and starving so they fed him one can of evaporated cream for breakfast, a can of sardines
for dinner and several slices of liver sausage for supper. Naturally the cat grew
strong and playful and much attached to its foster parents. One day a young seaman tried
to play with the kitten who promtly scratched him. Very much surprised the seaman
called out "That cat is crazy" The Carpenter indignantly cried "Sure, that cat is
crazy like Einstein. Wasn't he smart enough to leave Germany and come to the United
States?" So you have an eight weeks old kitten as a namesake. He has been formally,
respectfully and affectionately christened "Professor Albert Einstein " by a bunch
of hard boiled sailors whose conception of"relativity"is a euphemistic synonym for
"kinship"

Very respectfully,

Edward S.Moses,
Chief Engineer,
S.S.John M.Schofield,
Waterman S.S.Company,
19 Rector Street,
New York City.

August 10th,1946

Mr.Edward S.Moses,Chief Engineer
S.S.John M.Schofield
c/o Waterman S.S.Company
19 Rector Street
New York City

Dear Mr.Moses:

Thank you very much for your kind and
interesting information. I am sending my heartiest
greetings to my namesake, also from our own tomcat
who was very interested in the story and even a little
jealous. The reason is that his own name "Tiger" does
not express, as in your case, the close kinship to the
Einstein family.

With kind greetings to you, my namesake's
foster parents and to my namesake himself,

sincerely yours,

Albert Einstein.

ABOVE The chief engineer on the SS *John
M. Schofield* informs Einstein of a namesake
of his who happens to be a cat. Letter from
Edward S. Moses, at sea, August 3, 1946.

LEFT Einstein's humorous reply to the chief
engineer, August 10, 1946

OPPOSITE Albert and Elsa Einstein visit a set
at the Vitagraph film studios, Hollywood,
early 1930s
PHOTO COURTESY OF *LOOK* MAGAZINE

Einstein in his Princeton garden
with a delegation from the American
Continental Club, 1941

PHOTOS BY TRUDI DALLOS

OPPOSITE Princeton, 1950s

10.XII. Botschaften für Radio an Amerika
und für die zionistische Jugend geschrieben
Unzählige Telegramme, dass die Schiffsradioten

dem dicken Atlantis Schwarz.
Dazu ein Heer von Photographen
die sich wie ausgehungerte
Wölfe auf mich stürzten. Die
Reporter stellten ausgesucht blöde

10 | The Einstein Myth

"To punish me for my contempt for authority, fate made me an authority myself."

The Creation of the Myth

Albert Einstein is a powerful, mythical figure, a universal cultural icon. In popular mythology, he plays contrasting roles: saint and demon, genius and clown, wise old man and child, sorcerer and philosopher. There is a basic duality to the Einstein myth: Einstein is perceived as having unleashed the best and the worst in science and human nature.

Albert Einstein became famous overnight. Sensationalist reports of the verification of his general theory of relativity appeared in the international press on November 7, 1919. From then on, the press reported his every move and the general public became fascinated by this man who had overturned the traditional concepts of time and space. They were intrigued by his quirky appearance and quaint utterances. On his trips abroad, the masses flocked to see him.

Einstein's popularity was greatest among Jews: he was perceived as the embodiment of the Jewish intellect and the greatest Jew of his time. In reaction to this adulation, Einstein self-mockingly termed himself a "Jewish saint." Yet he was also demonized: anti-Semitic circles in Germany launched vitriolic attacks against him and his "Jewish physics."

Various factors contributed to the creation of the Einstein phenomenon: seemingly universal and absolute truths which had been valid for three hundred years were dramatically overturned by Einstein. A special mystique came to surround his allegedly incomprehensible theories. He was ascribed sorcerer-like qualities: he had discovered a magic formula—a key that opened the single secret

OPPOSITE The birth of the Einstein myth. Cover of the *Berliner Illustrirte Zeitung*, December 14, 1919. The caption reads: "A new celebrity in world history: Albert Einstein. His research signifies a complete revolution in our concepts of nature and is on a par with the insights of Copernicus, Kepler, and Newton."

© ULLSTEIN BILD

14. Dezember 1919

Nr. 50

28. Jahrgang

Berliner

Einzelpreis des Heftes

25 Pfg.

Illustrirte Zeitung

Verlag Ullstein & Co, Berlin SW 68

Phot. Swie Byk.

Eine neue Größe der Weltgeschichte: Albert Einstein,
dessen Forschungen eine völlige Umwälzung unserer Naturbetrachtung bedeuten
und den Erkenntnissen eines Kopernikus, Kepler und Newton gleichwertig sind.

Einstein surrounded by journalists, Pittsburgh, December 28, 1934
© THE NEW YORK TIMES

The cult of individuals is always, in my view, unjustified. To be sure, nature distributes her gifts unevenly among her children. But there are plenty of the well-endowed, thank God, and I am firmly convinced that most of them live quiet, unobtrusive lives. It strikes me as unfair, and even in bad taste, to select a few of them for boundless admiration, attributing super-human powers of mind and character to them. This has been my fate, and the contrast between the popular estimate of my powers and achievements and the reality is simply grotesque. The awareness of this strange state of affairs would be unbearable but for one pleasing consolation: it is a welcome symptom in an age which is commonly denounced as materialistic, that it makes heroes of men whose goals lie wholly in the intellectual and moral sphere. This proves that knowledge and justice are ranked above wealth and power by a large section of the human race.

—On the development of his popularity. From "My First Impression of the USA," July 1921.

of the universe. In addition, Einstein was seen as the personification of supreme intellect and his name became synonymous with genius.

After the First World War, a war-weary public was searching for a new kind of hero. Einstein was tailor-made for this role, with his anti-militarist views, bohemian appearance, gentle manner, and playful sense of humor.

In a wider cultural context, the arts, philosophy, and psychology were simultaneously creating their own revolutions in the approach to time and space: Picasso's cubism, Joyce's stream of consciousness, Schoenberg's atonal music, Bergson's concept of duration, and Freud's interpretation of dreams.

11.12: 7.30 AM. Arrival in New York. Was worse than my most fantastic expectations. Swarms of reporters boarded the ship at Long Island as well as the German Consul with fat, insufferable Schwarz. In addition, an army of photographers who pounced on me like starved wolves. The reporters asked really stupid questions to which I responded with cheap jokes which were greeted enthusiastically.

On being "mobbed" by the press upon his arrival in New York. From his travel diary to the United States, December 11, 1930.

Wherever I go and wherever I stay,
There's always a picture of me on display.
On top of the desk, or out in the hall,
Tied round a neck, or hung on the wall.

Women and men, they play a strange game,
Asking, beseeching: "Please sign your name."
From the erudite fellow they brook not a quibble
But firmly insist on a piece of his scribble.

Sometimes, surrounded by all this good cheer,
I'm puzzled by some of the things that I hear,
And wonder, my mind for a moment not hazy,
If I and not they could really be crazy.

—On the ubiquity of his image. Verse dedicated to Cornelia Wolff,
   January 1927. Translation: Helen Dukas and Banesh Hoffmann.

Schuster

Princeton, 4.5.1936

Liebe Nachwelt!

Wenn ihr nicht gerechter, friedlicher und überhaupt
vernünftiger sein werdet, als wir sind, bezw. gewesen sind, so
soll euch der Teufel holen.

Diesen frommen Wunsch mit aller Hochachtung geäussert
habend bin ich euer (ehemaliger)

gez.   Albert Einstein

Dear Posterity,
If you have not become more just,
more peaceful, and generally more
rational than we are (or were)—
why then, the Devil take you.
Having, with all respect, given
utterance to this pious wish,
I am (or was)
Yours,
Albert Einstein

Message to posterity written on parchment and placed in an air-tight metal box in the cornerstone of the house of the American publisher, M. L. Schuster, May 4, 1936

Albert and Elsa Einstein with Charlie Chaplin at the premiere of *City Lights*, Hollywood, January 30, 1931. In reaction to the crowds gaping at both of them in front of the theater, Chaplin remarked to Einstein: "The people applaud me because everyone understands me, and they applaud you because no-one understands you."

PHOTO BY EMIL HILB

The Myth in the Nuclear Age

With the onset of the atomic era after 1945, the Einstein myth underwent a radical transformation. Einstein was perceived as having created the threat of a nuclear holocaust—the new Prometheus who had brought atomic fire to humankind. Although this is a gross distortion of Einstein's role, it fulfilled the public's need to express its disillusionment with scientific and technological "progress." Einstein embodies the dilemma of the role of theoretical scientists in society: their theories describe universal laws of nature but they have no control over society's use of their knowledge.

Parallel to this process of demonization, Einstein became canonized as a saint. He was perceived as suffering on behalf of humanity—his soulsearching eyes, his crown of white hair, and his habit of not wearing socks further contributed to the image of a benevolent sage. Einstein's political campaigns on behalf of nuclear disarmament and civil liberties rendered him a moral authority for the post-war generation. Einstein's active commitment to societal change also had particular resonance among Jews: he became a symbol of Jewish morality and social conscience.

In the age of post-modernism, Einstein features prominently in newspaper articles, films, plays, operas, and advertisements. Yet his image has been depoliticized and trivialized: alongside Mickey Mouse and Marilyn Monroe, Einstein has become one of the major cultural icons of the modern era. The most recent example of Einstein's mythic status is his being named "Person of the Century" by *TIME* Magazine in 2000.

Anglo-Jewish sculptor Sir Jacob Epstein with his bust of Einstein. This artistic rendition of Einstein was created in the United Kingdom in September 1933, just prior to his emigration to the United States.

PHOTO BY THE ASSOCIATED PRESS

Albert Einstein with sculptor Gina Plunguian
working on her bust of Einstein, July 1947

Some examples of commercial use of Einstein's name and image

I have never given my name for commercial use even in cases when no misleading of the public was involved as it would be in your case. I, therefore, forbid you strictly to use my name in any way.

—From a letter to Mervin Ruebush, who asked Einstein for his permission to use Einstein's name to advertise his cure for stomach aches, May 22, 1942

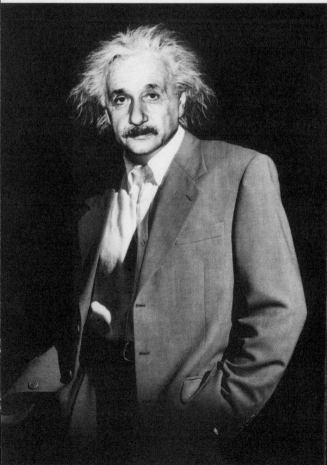

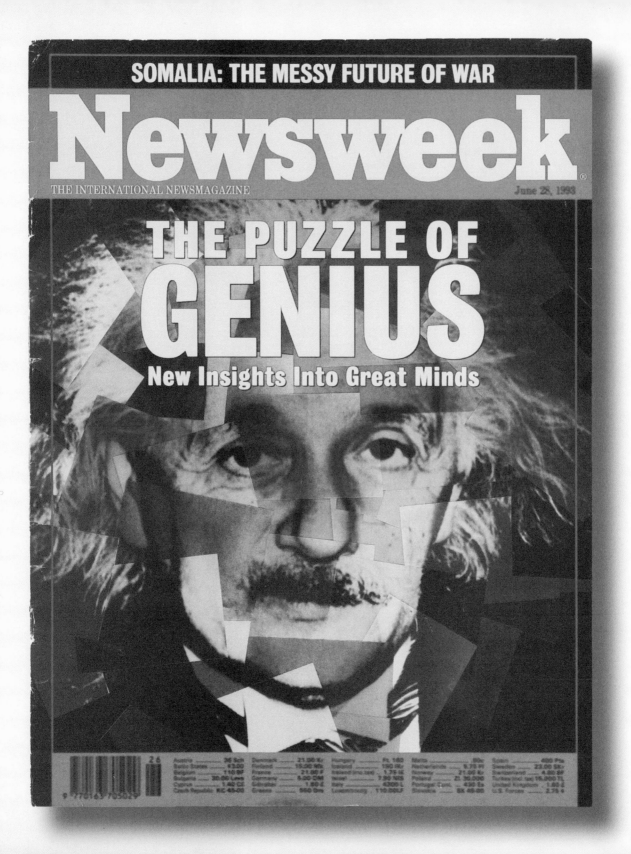

SOMALIA: THE MESSY FUTURE OF WAR

Newsweek

THE INTERNATIONAL NEWSMAGAZINE

June 28, 1993

THE PUZZLE OF
GENIUS

New Insights Into Great Minds

THE NEW YORK TIMES MAGAZINE

October 31, '13

RELATIVITY

Albert Einstein is universally associated with genius and personal nobility. He won the Nobel prize and worked for world peace. But according to "The Private Lives of Albert Einstein," just published in Britain, the gentle physicist was an adulterous, egomaniacal misogynist who may have even beaten his first wife. There is no question he once described her as a woman of "uncommon ugliness." Einstein himself was no Fabio, but his sagging jowls and unruly hair never seemed like Halloween material before. The newly revealed Bad Einstein, on the other hand, seems tailor made for trick or treating. Boo!

ABOVE LEFT The saint turns into the devil: publications about Einstein's private life created a backlash to his saintly image. Einstein as a Halloween mask.

© THE *NEW YORK TIMES MAGAZINE*, OCTOBER 31, 1993

ABOVE RIGHT The fate of Einstein's dissected brain continues to fascinate the public's imagination. Its remains are preserved in a jar in Princeton.

PHOTO BY DAVID HUTSON

OPPOSITE On the cover of *Newsweek,* June 28, 1993

© 1993 NEWSWEEK, INC. / J. WOLFF / BETTMANN CORBIS; ALL RIGHTS RESERVED; REPRINTED BY PERMISSION

The Albert Einstein Archives

"My final object is that any such property which may remain (whether it consist of original manuscripts, or literary rights or property still owned by the estate) shall pass to The Hebrew University and become its property absolutely, to be thereafter retained or disposed of by it as it may deem to be in its best interests."

—The Last Will and Testament of Albert Einstein
March 18, 1950, par. 13

History of the Archives

Albert Einstein was not the sort of person to retain every piece of paper that passed through his hands. He made no systematic attempt to preserve his papers prior to 1919. As a result of his dramatic rise to fame in November 1919, his correspondence increased vastly and he employed his stepdaughter, Ilse Loewenthal, as his first secretarial assistant. She achieved the first semblance of well-ordered files. In April 1928, Helen Dukas came to work for Einstein and began to preserve his papers more systematically. However, not even then were copies of all outgoing correspondence kept. Shortly after the Nazis' rise to power in 1933, Einstein's papers were rescued from Berlin by Einstein's son-in-law, Rudolf Kayser, with the help of the French Embassy. The material was brought to Einstein's new home in Princeton and kept there until well after his death. With a few exceptions, the material left at Einstein's summer house in Caputh outside Berlin was destroyed in order to prevent it falling into the hands of the Nazi authorities.

Einstein's will of 1950 appointed his secretary, Helen Dukas, and his close associate, Otto Nathan, as trustees of his estate. Nathan was also appointed as sole executor. Following Einstein's death in 1955, Dukas and Nathan devoted themselves tirelessly for a quarter of a century to organizing the papers and acquiring additional material. As a result of their efforts, the Archives grew threefold. In the 1960s, Helen Dukas and Prof. Gerald Holton of Harvard University reorganized the material, thereby rendering it accessible to scholars and preparing it for eventual publication in *The Collected Papers of Albert Einstein,* a joint project of The Hebrew University and Princeton University Press. To facilitate editorial work, the papers were transferred from Einstein's home to the Institute for Advanced Study in Princeton. In 1982, the Einstein Estate transferred all literary rights to The Hebrew University and Einstein's personal papers were transferred to The Jewish National & University Library in Jerusalem. President Avraham Harman of The Hebrew University and Prof. Milton Handler of the American Friends of The Hebrew University played a crucial role in securing the transfer of the material to Jerusalem. In subsequent years, additional material was dispatched from Einstein's Princeton residence, namely his personal collections of reprints, photographs, medals, and diplomas as well as his private library. In 1988, the Bern Dibner Curatorship for the running of the Albert Einstein Archives was established by the Dibner Fund of Connecticut.

The Importance of the Archives

The Albert Einstein Archives is an extraordinary cultural asset of universal importance for humanity and of national importance for Israel and the Jewish people. Representing the intellectual and personal record of a creative genius whose thinking profoundly changed our perception of the universe, it is of inestimable value. Einstein did not wish that any physical monument or memorial be erected in his name. The preservation of his papers, which most authentically reflect his ideas and person, affords a far more fitting means of maintaining his legacy.

The Albert Einstein Archives contains the largest collection of original manuscripts by Einstein in the world and includes his vast correspondence with the most influential physicists and intellectuals of the twentieth century. Moreover, it comprises the most exhaustive compilation of material about Albert Einstein.

The Albert Einstein Archives constitutes an extremely valuable historical resource. It is considered one of the most significant sources for the history of modern physics. In addition, the Archives is an extremely important source for the history of such movements as pacifism, socialism, and Zionism as well as for German, Jewish, European, and American intellectual, political, and social history of the twentieth century.

Bibliography

Primary Literature

Boni, Nell, Monique Russ, and Dan H. Laurence. *A Bibliographical Checklist and Index to the Collected Writings of Albert Einstein* (New York: Readex Microprint Corporation, 1960).

Calaprice, Alice. *The Quotable Einstein* (Princeton, New Jersey: Princeton University Press, 1996).

Dukas, Helen, and Banesh Hoffmann, eds. *Albert Einstein: The Human Side. New Glimpses from His Archives* (Princeton, New Jersey: Princeton University Press, 1979).

Einstein, Albert. *About Zionism* (London: The Soncino Press, 1930).

———. *Ideas and Opinions* (New York: Crown Publishers, Inc., 1982).

———. *The Meaning of Relativity* (London: Chapman and Hall, 1991).

———. *Out of My Later Years* (New York: The Philosophical Library, Inc., 1950).

———. *Relativity* (London: Methuen and Co., Ltd., 1988).

———. *The World As I See It* (New York: Carol Publishing Group, 1991).

Einstein, Albert, and Michele Besso. *Correspondence 1903–1955* (Paris: Hermann, 1972).

Einstein, Albert, and Max Born. *The Born-Einstein Letters* (London: The Macmillan Press Ltd., 1971).

Einstein, Albert, and Sigmund Freud. *Why War?* (Chicago: Chicago Institute for Psychoanalysis, 1978).

Einstein, Albert, and Leopold Infeld. *The Evolution of Physics* (New York: Simon and Schuster, 1938).

Klein, Martin J., A. J. Kox, Jürgen Renn, and Robert Schulmann, eds. *The Collected Papers of Albert Einstein*, vol. 3, *The Swiss Years: Writings, 1909–1911* (Princeton, New Jersey: Princeton University Press, 1993).

———. *The Collected Papers of Albert Einstein*, vol. 4, *The Swiss Years: Writings, 1912–1914* (Princeton, New Jersey: Princeton University Press, 1995).

———. *The Collected Papers of Albert Einstein*, vol. 5, *The Swiss Years: Correspondence, 1902–1914* (Princeton, New Jersey: Princeton University Press, 1993).

Lief, Alfred, ed. *Albert Einstein: The Fight Against War* (New York: The John Day Company Inc., 1938).

Nathan, Otto, and Heinz Norden, eds. *Einstein on Peace* (New York: Simon and Schuster Inc., 1960).

Renn, Jürgen, and Robert Schulmann, eds. *Albert Einstein and Mileva Marić: The Love Letters* (Princeton, New Jersey: Princeton University Press, 1992).

Schilpp, Paul Arthur, ed. *Albert Einstein: Autobiographical Notes* (La Salle, Illinois: Open Court Publishing Company, 1979).

Stachel, John, ed. *The Collected Papers of Albert Einstein*, vol. 1, *The Early Years: 1879–1902* (Princeton, New Jersey: Princeton University Press, 1987).

———. *The Collected Papers of Albert Einstein*, vol. 2, *The Swiss Years: Writings, 1900–1909* (Princeton, New Jersey: Princeton University Press, 1989).

Secondary Literature

Aldridge, Susan. "Relative Values," *New Statesman and Society* (September 3, 1993): 37–38.

Bergmann, Peter G. "There Are No Miracles," *The Sciences* (March 1979): 2–3, 30–31.

Bernstein, Jeremy. *Albert Einstein and the Frontiers of Physics* (New York: Oxford University Press, 1996).

Brian, Denis. *Einstein: A Life* (New York: John Wiley & Sons, Inc., 1996).

Bucky, Peter A. (with the collaboration of Allen G. Weakland). *The Private Albert Einstein* (Kansas City: Andrews and McMeel, 1992).

Clark, Ronald W. *Einstein: The Life and Times* (London: Hodder and Stoughton, 1973).

Davies, P. C. W. "Einstein's Legacy," *The Sciences* (March 1979): 25–28.

Fölsing, Albrecht. *Albert Einstein: Eine Biographie* (Frankfurt am Main: Suhrkamp Verlag, 1993).

Forward, Robert L. "Einstein's Legacy," *OMNI* (1979): 54–59.

Frank, Philipp. *Einstein: His Life and Times* (New York: Da Capo Press, Inc., 1947).

French, A. P., ed. *Einstein: A Centenary Volume* (Cambridge, Massachusetts: Harvard University Press, 1979).

Friedman, Alan J., and Carol C. Donley. *Einstein as Myth and Muse* (Cambridge: Cambridge University Press, 1985).

Goldsmith, Maurice, Alan Mackay, and James Woudhuysen, eds. *Einstein: The First Hundred Years* (Oxford: Pergamon Press, 1980).

Guterl, Fred. "Keyhole View of a Genius," *Scientific American* (January 1994): 15–16.

Highfield, Roger, and Paul Carter. *The Private Lives of Albert Einstein* (London: Faber and Faber, 1993).

Hoffmann, Banesh (with the collaboration of Helen Dukas). *Albert Einstein: Creator and Rebel* (New York: Viking Press, 1972).

Holton, Gerald. "Of Love, Physics and Other Passions: The Letters of Albert and Mileva," *Physics Today* (August 1994): 23–29 and (September 1994): 37–43.

Holton, Gerald, and Yehuda Elkana, eds. *Albert Einstein: Historical and Cultural Perspectives. The Centennial Symposium in Jerusalem* (Princeton, New Jersey: Princeton University Press, 1982).

Jammer, Max. *Einstein und die Religion* (Konstanz: Universitätsverlag Konstanz, 1995).

Jewish National & University Library. *Einstein: 1879–1979 Exhibition* (Jerusalem: Raphael Haim Hacohen Press Ltd., 1979).

Kantha, Sachi Sri. *An Einstein Dictionary* (Westport, Connecticut: Greenwood Press, 1996).

Klein, Martin J. "A Bit of Light," *The Sciences* (March 1979): 10–13.

Klein, Martin J., A. J. Kox, and Robert Schulmann, eds. "Introduction to Volume 5," in *The Collected Papers of Albert Einstein*, vol. 5 (Princeton: Princeton University Press, 1993).

Lask, Thomas. "World of the Star and the Atom," *New York Times* (November 4, 1972).

Lewis, Albert C. *Albert Einstein, 1879–1955: A Centenary Exhibit of Manuscripts, Books, and Portraits* (Austin: The Humanities Research Center, The University of Texas, 1979).

Nathan, Otto, and Heinz Norden, eds. "Vorbemerkungen der Herausgeber," in *Albert Einstein: Über den Frieden. Weltordnung oder Weltuntergang?* (Bern: Herbert Lang and Cie AG, 1975).

National Museum of American History. *Einstein: A Centenary Exhibition at the*

National Museum of History and Technology (Washington, D.C.: Smithsonian Institution Press, 1979).

Neumann, Thomas, ed. *Zeitmontage: Albert Einstein* (Berlin: Elefanten Press, 1989).

Pais, Abraham. *Einstein Lived Here* (New York: Oxford University Press, 1994).

———. *"Subtle is the Lord . . .": The Science and the Life of Albert Einstein* (New York: Oxford University Press, 1982).

Penrose, Roger. "Einstein's Vision and the Mathematics of the Natural World," *The Sciences* (March 1979): 6–9.

Renn, Jürgen, and Robert Schulmann, eds. "Introduction," in *Albert Einstein and Mileva Marić: The Love Letters* (Princeton, New Jersey: Princeton University Press, 1992).

Resnick, Robert. "Misconceptions about Einstein: His Work and His Views," *Journal of Chemical Education* 57 (December 1980): 854–62.

Sayen, Jamie. *Einstein in America* (New York: Crown Publishers, Inc., 1985).

Stachel, John. "Albert Einstein: The Man Beyond the Myth," *Bostonia* (February 1982): 9–17.

———. "Einstein's Odyssey," *The Sciences* (March 1979): 14–15, 32–34.

Stachel, John, ed. "General Introduction," in *The Collected Papers of Albert Einstein*, vol. 1 (Princeton, New Jersey: Princeton University Press, 1987).

Stent, Gunther S. "Does God Play Dice?" *The Sciences* (March 1979): 18–23.

Sugimoto, Kenji. *Albert Einstein: A Photographic Biography* (New York: Schocken Books, 1989).

Warnow, Joan N., and the American Institute of Physics. *Images of Einstein: A Catalog* (New York: American Institute of Physics, 1979).

White, Michael, and John Gribbin. *Einstein: A Life in Science* (London: Simon and Schuster, 1993).

Woolf, Harry, ed. *Some Strangeness in the Proportion: A Centennial Symposium to Celebrate the Achievements of Albert Einstein* (Reading, Massachusetts: Addison-Wesley Publishing Company, 1980).

"The Year of Dr. Einstein," *Time* (February 19, 1979): 70–79.

Index

civil rights, 80, 178
cold war, 80, 138
The Collected Papers of Albert Einstein,
    186
Columbia University, 116
Committee on Intellectual Cooperation
    (League of Nations), 4, 66–67
communism, 84. *See also* McCarthyism;
    socialism
Conference of Jewish Students
    (Germany), 96–97
conscientious objection, 66
Copernican Quadricentennial (1943;
    Carnegie Hall, New York), 129
Copernicus, 172
cosmological constant, revision of
    theory of, 129

Damman, Ruth, 147
Daughters of the American Revolution,
    82
democracy, 66, 79, 111
Dewey, John, 129
Dhomme, Father, 93
disarmament. *See* arms control
Disney, Walt, 129
Dizengoff, Meir, 93
Dodds, H. W., 125
Donahue, Gerald M., 90
Dukas, Helen, 5, 17–18, 21, 84, 137, 176;
    and Einstein archives, 185–86

Eban, Abba, 103
Eddington, Arthur, 41
Edison, Thomas Alva, 131
education, 84–85; of AE, 10–11, 28
Ehrenfest, Paul, 41, 47, 118

Einstein, Albert (AE): *annus mirabilis*
    (1905) of, 2, 28, 30; anti-war
    activities of, 66, 76–77, 101; archives
    of, 185–87; Berlin years of, 3–4, 36;
    birth of, 2, 8; brain of, 183; in
    California, 119–23; at Caltech, 61,
    132–34; childhood of, 2; commercial
    use of name, 180–81; correspon-
    dence with children, 149–57;
    correspondence with scientists,
    47–50, 66–69, 76–77, 128–29, 131,
    134, 187; curiosity file of, 159–70;
    death of, 5, 21, 185; demonization of,
    172, 178, 183; divorce of, 3; and dog
    Chico, 21, 23; early years in science
    of, 28–31; on education, 84–85;
    education of, 10–11, 28; family life
    of, 12–20; in Far East, 4, 15; and FBI,
    80, 111; first four papers (1905) of,
    30–31; on freedom of inquiry, 60; on
    God, 89; health of, 3, 5; Jewish
    identity of, 87–108, 172, 178; on the
    Jewish people, 94, 96, 136, 138; at
    leisure, 140–47; marriage of (first),
    2, 12–13, 18, 22; marriage of (second),
    3, 18, 22–23; message to posterity
    from, 177; and music, 142–45; myth
    of, 172–78; naturalization of, 18–19;
    need for solitude of, 7, 23; pacifism
    of, 66–71; personality of, 22–23;
    "plumber statement" of, 82; poems
    by, 113, 177; political activity of,
    65–85; popularity of, 172, 174–77;
    Princeton years of, 4–5, 18–21,
    54–59, 72, 124, 126–27, 168–69; and
    psychotherapy, 23; scientific achieve-
    ments of, 25–64; on scientists and